W9-BYV-962

BIODIVERSITY IN ECOSYSTEMS

Tasks for vegetation science 34

SERIES EDITORS

The titles published in this series are listed at the end of this volume.

Biodiversity in ecosystems: principles and case studies of different complexity levels

Edited by

A. KRATOCHWIL

KLUWER ACADEMIC PUBLISHERS

DORDRECHT/BOSTON/LONDON

Library of Congress Cataloging-in-Publication Data.

ISBN 0-7923-5717-5

Published by Kluwer Academic Publishers,
P.O. Box 17, 3300 AA Dordrecht, The Netherlands.

Sold and distributed in North, Central and South America
by Kluwer Academic Publishers,
101 Philip Drive, Norwell, MA 02061, U.S.A.

In all other countries, sold and distributed
by Kluwer Academic Publishers,
P.O. Box 322, 3300 AH Dordrecht, The Netherlands.

Printed on acid-free paper

Printed in the Netherlands.

Contents

Editorial

The subject of biological diversity is a highly complex one; its treatment must accordingly be comprehensive and varied. This volume does not aim at merely adding to the immensely grown and meanwhile apparently incalculable number of individual publications on "biodiversity", all the more since already many excellent articles and books have been published (see *e.g.* List of References Kratochwil in this volume). It is rather our objective to investigate biodiversity on the up to now little studied coenosis and landscape levels. Phytosociological and animal-ecological fields are considered, as well as theoretical approaches to biodiversity and aspects of its application in nature and landscape protection and preservation. Since biodiversity has so far been predominantly studied in the Anglo-American area, it seemed to be of value to discuss this complex topic from a Central and southern European viewpoint, based on data gathered in these regions, and to thus promote a global discussion.

The volume "Biodiversity in ecosystems: principles and case studies of different complexity levels" comprises twelve publications, divided into three chapters:
1) "Biodiversity: survey and principles"
2) "Fauna, flora, and vegetation in ecosystems: some aspects of biodiversity"
3) "Biodiversity and nature preservation"

In order to further develop scientific theories, it is first of all necessary to synoptically depict the structure of the theory with its individual principles, concepts, and hypotheses. This is done in chapter 1: "Biodiversity: survey and principles." A definition of the concept "biodiversity" and its ranges of validity are presented, as well as different forms of intra- and interbiocoenotic diversity. Special emphasis lies on the formulation of certain questions and hypotheses on biodiversity. It is not so much the authors' objective to work out generalities, but to compile - and, if possible, specify - general statements on biodiversity always recurring in the literature. While in A. Kratochwil's article 30 hypotheses on biodiversity are discussed, in most cases critically and including counter-hypotheses, M. Schaefer's contribution shows, at models and examples, the direct linkage of concrete results of a 15-year research project on Central European beech forest ecosystems to important hypotheses on biodiversity.

The major part of the articles is found in chapter 2: "Fauna, flora, and vegetation in ecosystems: some aspects of biodiversity." In this chapter, single case studies are presented, encompassing quite different ecological objects, spatial dimensions, and complexity degrees. G. and S. Pignatti's article deals with Mediterranean ecosystems and the spatial distribution of species diversity. Small-scale mapping of biodiversity will in future be especially important to find solutions to global biodiversity problems. One great challenge will be to depict the biodiversity of the earth's ecosystems within the framework of geographical information systems. The contribution demonstrates in an impressing manner coincidences of the plant species diversity in Italy with the macroclimate (temperature, precipitation etc.). The authors emphasize the importance of human impact on biodiversity, without which the species diversity in the Mediterranean area would be much lower, compared to the vegetation under natural conditions.

A. Kratochwil (ed.), Biodiversity in ecosystems, 1-3

Whereas G. and S. Pignatti's article focuses on a small-scale treatment of "biodiversity", A. Schwabe's work aims at its registration and analysis on a large scale. In this context, sigmasociology constitutes an especially valuable tool. By the recording and analysis of vegetation complexes, considering also microhabitats, landscape units are differentiated on a phytosociological basis. A landscape mosaic can thus be typified. Case studies have shown that the method is broadly applicable; also large areas can be precisely recorded and analyzed as to their phytocoenotic diversity. By determining different degrees of anthropogenic influence (hemeroby), the influence of man on biodiversity can be "measured". The applicability of this method is of special importance for Environmental Impact Assessments.

R. Pott's article on the pasture-woodlands in north-western Germany shows exemplarily that the great biodiversity in Central Europe is due to extensive management by man since the Neolithic Period. The diversity of different vegetation types with the biocoenoses they are composed of reflects a broad repertoire of varying historical management forms. A preservation of such natural and cultural landscapes presupposes comprehensive knowledge about their origin and development.

The article by V.K. Brown and A.C. Gange demonstrates at models that there is a direct causal connection between plant diversity and insect herbivory. A causal analysis of biodiversity therefore requires the study of biocoenoses. The development of a biocoenological structure in the course of succession is an interactive process, during which leaf-eating insects increase the plant diversity by altering the competitive situation and by creating microhabitats.

Similar to the contribution by Pignatti and Pignatti, H. Mattes chose a small-scale approach to assess biodiversity. In his article, he deals with the biogeography and the species diversity of bird communities of coniferous forests in Eurasia. Here, too, certain diversity centres have been found, which however cannot solely be explained by currently effective ecofactors. Fauna-historical aspects are the key to special diversity patterns.

The calculation of diversity indices is still a common procedure to compare the diversities of different localities or habitats. R. Schröpfer shows exemplarily the limitations of such procedures, *e.g.* for mammal communities, in which only few dominant species characterize the species community.

One basic question is whether deterministic or stochastic relations affect the degree of biodiversity at all. J.U. Ganzhorn demonstrates at the example of Malagasy lemurs that their diversity depends on the quality of vegetation, but also on certain historical influences on the lemur community. Thus always a wide range of different (in this case, as it can be proven, deterministic) relations has to be taken into account when analyzing a diversity pattern.

The third and final chapter is about biodiversity and nature preservation. W. Haber outlines fundamental aspects of the concept "biodiversity" and its study, and problems arising in this context. The author warns of false hopes: his critical article shows scientific, social, and political limitations. H. Haeupler's contribution demonstrates that, due to inadequate basic scientific knowledge, it is very difficult to develop pragmatic approaches to the - more and more pressing - issue of protection and preservation of biodiversity. The solution can only be an intensification of research, putting - as it is currently done - special emphasis on the level of population ecology, since here the lack of knowledge is considerable.

The volume concludes with a contribution of the Federal Agency for Nature Conservation (Institute for Biotope Protection and Landscape Ecology) by J. Blab, M. Klein, and A. Ssymank, presenting the scope within which measures can be taken, and the prospects for the future, from the point of view of a German federal authority. The authors show the linkage of basic scientific knowledge, objective and realization of a concept "biodiversity". Within Europe, the legal framework is provided by the habitat-directive of the European Union.

The treatment of each topical subject is important, independent of whether it consolidates known hypotheses and theories or alters a paradigm. Much more basic research is required for the development of a comprehensive and detailed "general theory of biodiversity." The danger, however, that the gulf between theory and practice is becoming too great, is increasing. Some current concepts, hypotheses, and theories take, as the examples of the "island theory" and the "metapopulation concept" show, the following course: First the relations are discussed by theorists, ingeniously and in great detail, usually defining the limitations and the prevailing conditions. Then the concepts developed are translated into practical measures (in this case of nature protection and preservation), and become paradigms; they are frequently misunderstood and overinterpreted. Often the theoretical generalizations in the field of applied nature protection and preservation are too imprecise, too superficial and too much simplified to allow predictions, and they can hardly be proved. Many concepts are raised to dogmata, without doing justice to the diversity of the ecological and historical conditions of biological systems. May this not apply to the "theory of biodiversity".

We would like to thank Kluwer Academic Publishers for their support in printing this volume, and Ms. Martina Lemme, University of Osnabrück, Department of Ecology, for having translated the articles by A. Kratochwil, R. Pott, and W. Haber, as well as this Editorial, into English, and for linguistic revision of the remaining contributions (except for the ones by A. Schwabe-Kratochwil and V.K. Brown).

A. Kratochwil, Osnabrück

CHAPTER 1

BIODIVERSITY SURVEY AND PRINCIPLES

BIODIVERSITY IN ECOSYSTEMS: SOME PRINCIPLES

ANSELM KRATOCHWIL

Fachgebiet Ökologie, Fachbereich Biologie/Chemie, Universität Osnabrück, Barbarastraße 11, D-49069 Osnabrück, Germany

Keywords: biocoenology, biodiversity, community ecology, community research, structural and functional diversity

Abstract

A scientific treatment of biodiversity must aim at the development of a generally valid theory, for only in this way, the scientific foundations for pragmatic approaches to the preservation of biodiversity can be worked out. This is also embodied in the "International Convention" on the protection and conservation of biodiversity, which was passed at the United Nations' conference "Environment and Development" at Rio de Janeiro in 1992, and has meanwhile been ratified by numerous nations. In this article, a definition of the concept "biodiversity" will be given, and partly synonymously used terms, like "variation", "differentiation", "diversification", "heterogeneity", "variety", "variability", "complexity", and "richness" be differentiated. As matter and energy occur in very different organization forms and on varying hierarchy levels, it is necessary to delimit first those areas in which biodiversity phenomena appear.

An essential prerequisite for a scientific investigation of the diversity of biocoenoses and ecosystems is the theoretical examination of opposing positions: holistic approach versus individualistic approach, deterministic approach versus stochastic approach. Central questions are studied, concerning also ecology in general.

The varying forms of biodiversity can basically be assigned to four different groups: diversity of elements (element pattern of biodiversity), diversity of interactions (dynamic pattern of biodiversity), mechanisms causing diversity (causing pattern of biodiversity), and process of functioning (functional pattern of biodiversity). Examples will be given for each group. "Intrabiocoenotic diversity" includes the diversity of phytocoenoses, zoocoenoses, and synusia. A classification into synusia should

A. Kratochwil (ed.), Biodiversity in ecosystems, 5-38

follow stratotope, choriotope, and merotope patterns. "Interbiocoenotic diversity", on the other hand, refers to the diversity of landscape parts (vegetation complexes) and landscapes (vegetation complexes occurring together). Some fundamental relations will be outlined.

On the basis of the comprehensive discussion of biodiversity criteria in the literature, 30 hypotheses on biodiversity will be presented and explained. A final chapter deals with the importance of the theory of biodiversity in applied nature protection. The preservation of biodiversity is a basic component of the survival programme "sustainable development".

Introduction

Experts, but also the general public, agree that the preservation of biodiversity is of paramount importance. Accordingly, on the United Nations' conference "Environment and Development", held at Rio de Janeiro in 1992, an official "International Convention" on the protection and conservation of biodiversity was drawn up and, within two years, ratified by the majority of the undersigning nations. This official declaration of intent obliges the international "scientific community" to work out the scientific principles of a "theory of biodiversity" (Solbrig 1991); an important contribution to this has to be made by ecologists and biocoenologists/community ecologists (Raustiala & Victor 1996; Haber in this volume). Numerous, partly voluminous treatises give evidence of the particular relevance of biodiversity, and of the general efforts to do justice to its complexity (Wilson 1988, 1992; During *et al.* 1988; Stearns 1990; McNeely *et al.* 1990; Solbrig 1991, 1994; Solbrig *et al.* 1992; Courrier 1992; Groombridge 1992; World Resources Institute 1992; Schulze & Mooney 1993; Ricklefs & Schluter 1993; UNEP 1993, 1995; Huston 1994; Krattinger *et al.* 1994; Heywood & Watson 1995; Kim & Weaver 1995; Rosenzweig 1995; Haeupler 1997; Reaka-Kudla *et al.* 1997).

The objective of a scientific study of biodiversity is the development of a general theory (Fig. 1).

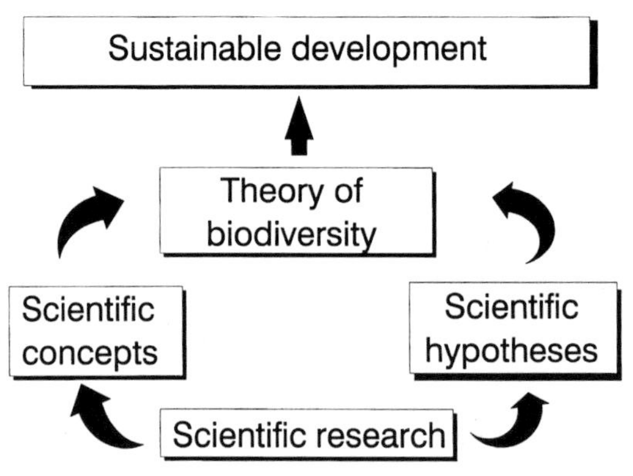

Figure 1. Ways to the formulation of a "theory of biodiversity", to be incorporated into the concept of a "sustainable development".

To this end, scientific data are gathered from which natural laws shall be deduced. The formation of a theory requires the formulation of scientific concepts and hypotheses. If there is evidence for general principles, it is possible to make prognoses and to create scenarios. In order to successfully cope with future tasks in accordance with the programme "sustainable development", predictions on the consequences of altered environmental conditions and the development of scenarios are of considerable significance (Kratochwil 1996). One of the key issues is the importance of the preservation of biodiversity for the maintenance of the global natural balance. Apart from the question to what degree it should be preserved, it has first and foremost to be studied to what extent it can be preserved (Blab *et al.* 1995).

To find out whether biodiversity is governed by certain natural laws is only one component of the analysis. Since the historicity of life is one of the fundamental characteristics of biology and thus also of ecology (Whittaker 1972; Osche 1975), evolution-biological and evolution-ecological aspects have to be considered in concepts and hypotheses, and have to be incorporated into a general theory of biodiversity. Hence follows that not all forms of biodiversity are repeatable at any time and any place. Dollo's law on the irreversibility of evolution-historical processes implies that a plant or animal species can only originate once. Structures lost in the course of evolution can never be regained in their original form (see *e.g.* Osche 1966). What is lost, is irretrievably lost, since evolution - as genealogical process - takes a linear course, not a cyclic one.

When investigating biodiversity, man has to be considered as biological and ecological factor, too. A study of biodiversity thus allows not only a more detailed understanding of life processes in general, but also of human life; moreover it is, as component of the survival programme "sustainable development", important for the future of man (Fig. 2).

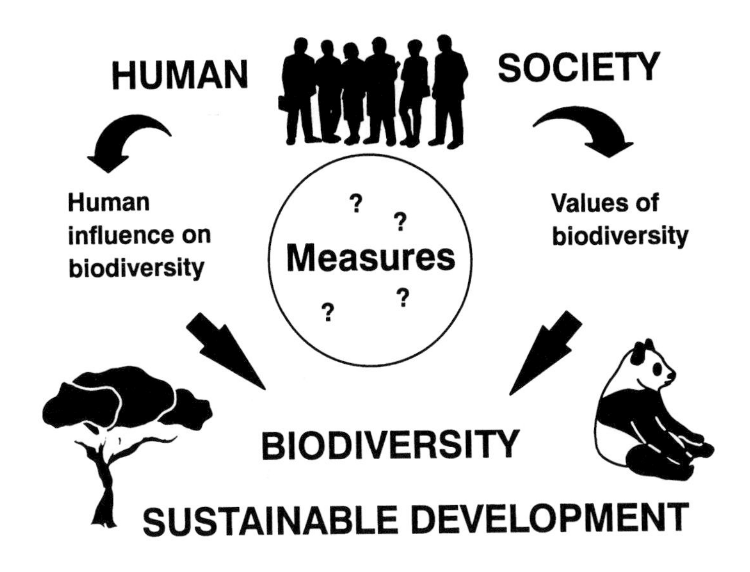

Figure 2. The interaction between human society and biodiversity (based on Heywood & Baste 1995).

Like biodiversity is closely linked with biological evolution, cultural diversity and the cultural and historical development of mankind (cultural evolution) go hand in hand (Gadgil 1987). For both, the many and diverse processes are system-inherent. Losses of their material and non-material values are always irrecoverable. The question which biological and cultural "achievements" should be preserved can only be answered by individual assessments.

Two central issues have to be investigated first:

- What is biodiversity? (see Akeroyd 1996)
- How can biodiversity be measured? (see *e.g.* Whittaker 1972; Magurran 1988; Hawksworth 1995).

What is biodiversity?

In its original sense, diversity means "variation" and "differentiation", "diversification", in contrast to "uniformity". Diversity may be understood as something static: "heterogeneity" then denotes irregularities, "variety" differences. "Variability" covers dynamic aspects. Diverse systems may be simple, but also very complicated. As a rule, complexity is a sure sign of diverse systems: it is defined as something very intricate or complicated. Complexity covers the profundity of system structures, diversity their width. When assessing biological systems, diversity may also be seen as "richness".

By biodiversity, biological diversity is understood: the total differentiation, variation, variability, complexity, and richness of life on earth.

These definitions show already the catch in such "condensed" terms: their ambiguity, which may help a layman to associate a number of things with a term (when being given the relevant information), but may of course also lead to misinterpretations; that is why experts avoid using merely the terms, but attempt to more closely define and differentiate the concepts behind (Akeroyd 1996; Haber in this volume). Many biological and ecological terms (and concepts) have been subject to a similar development: the concept "ecology" itself, the "island theory", and the "metapopulation concept" (Kratochwil 1998). The problem of the complexity and ambiguity of concepts can hardly be solved; they have always been renamed and redefined, and will also be in future. All the same, concepts must be defined as "tools" and as means to exchange information, in accordance with a general convention.

The following definition of biodiversity is proposed (based on art. 2 of the "Convention on Biological Diversity" of the IUCN, Rio de Janeiro 1992, altered after Bisby 1995): " 'Biological diversity' means diversity (according to differentiation, variation, variability, complexity and richness) among living organisms from all sources, including, inter alia, terrestrial, marine and other aquatic ecosystems and the ecological complexes of which they are part, this includes diversity within species, between species and of ecosystems".

Ranges of validity of biodiversity

Diversity is a fundamental quality, manifesting itself in the different organization levels of matter and energy (Haber 1978; Odum 1983). It is a characteristic feature of all levels of the non-biological and the biological hierarchy (hierarchical diversities) (Fig. 3); there is diversity on every single level. The levels of life are particularly diverse, here we generally distinguish between structural diversities and functional diversities (Solbrig 1991).

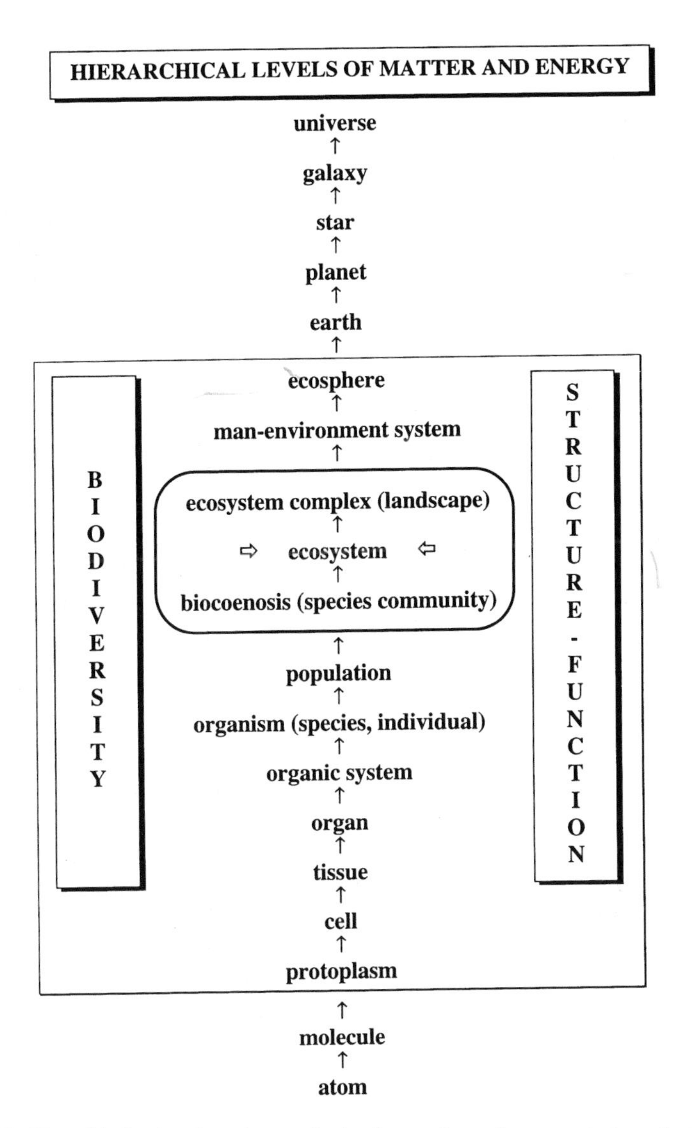

Figure 3. The hierarchical order of certain organization forms of organisms or parts thereof, of matter and energy (altered after Haber 1978), and the rank of biocoenology within the different organization levels.

Data on diversity may be studied at each level of the hierarchical structure, using two different approaches:
- a descriptive approach (*e.g.* identification, determination, description and differentiation of elements and their components);
- a functional approach (*e.g.* a causal analysis of the combination of the elements and their components, as well as of absorption, transformation, and processing of energy and matter).

Subject of this analysis is the level of ecosystems in the broader sense: their biotic components (biocoenoses) and their habitats (biotopes). Moreover the level of ecosystem complexes (landscape units) will be dealt with. Such complexes are formed by several ecosystems, the correlations of which follow certain rules. Since the Neolithic Period and increasingly in the past 150 years, man has considerably influenced ecosystems and ecosystem complexes in many parts of the world. A study of biodiversity therefore must include "man-environment systems". An increase in, but also a reduction of biodiversity may be anthropogenically caused.

On most organization levels of matter and energy (Fig. 3), the objects are supposed to represent entities. This, however, is not generally accepted for the levels of biocoenoses and ecosystems.

On the scientific treatment of biodiversity at the level of biocoenoses, ecosystems, and ecosystem complexes

An essential prerequisite for a scientific investigation of biocoenoses and ecosystems is the intensive theoretical examination of several, widely diverging approaches. The discussion focuses on two different viewpoints (Fig. 4) (after Trepl 1988, 1994):
a) holistic approach versus individualistic approach
b) deterministic approach versus stochastic approach.

There are gradual transitions between those widely diverging viewpoints. The two extreme views "ecosystems as super-organisms" or "as mere by-products" (the latter designated as "Gleasonian approach", Gleason 1926) are not endorsed by many. The majority of scientists rather follows deterministic or functionally based approaches (after Elton 1933). Opinion is divided as to the assessment of random events as system component. A deterministic principle does not necessarily exclude such events in a certain phase. In the course of succession of vegetation, an early stage may largely depend on random colonization, a latter one to a lesser extent. It seems pointless to analyze the importance of determinism, of stochastics, or of probability without examples and without relation to concrete objects, since in nature there is less an "either - or" than rather a "both - and" of phenomena.

It is agreed that biodiversity must not be seen as purely static, but that life on the different hierarchy and complexity levels always implies a dynamic component.

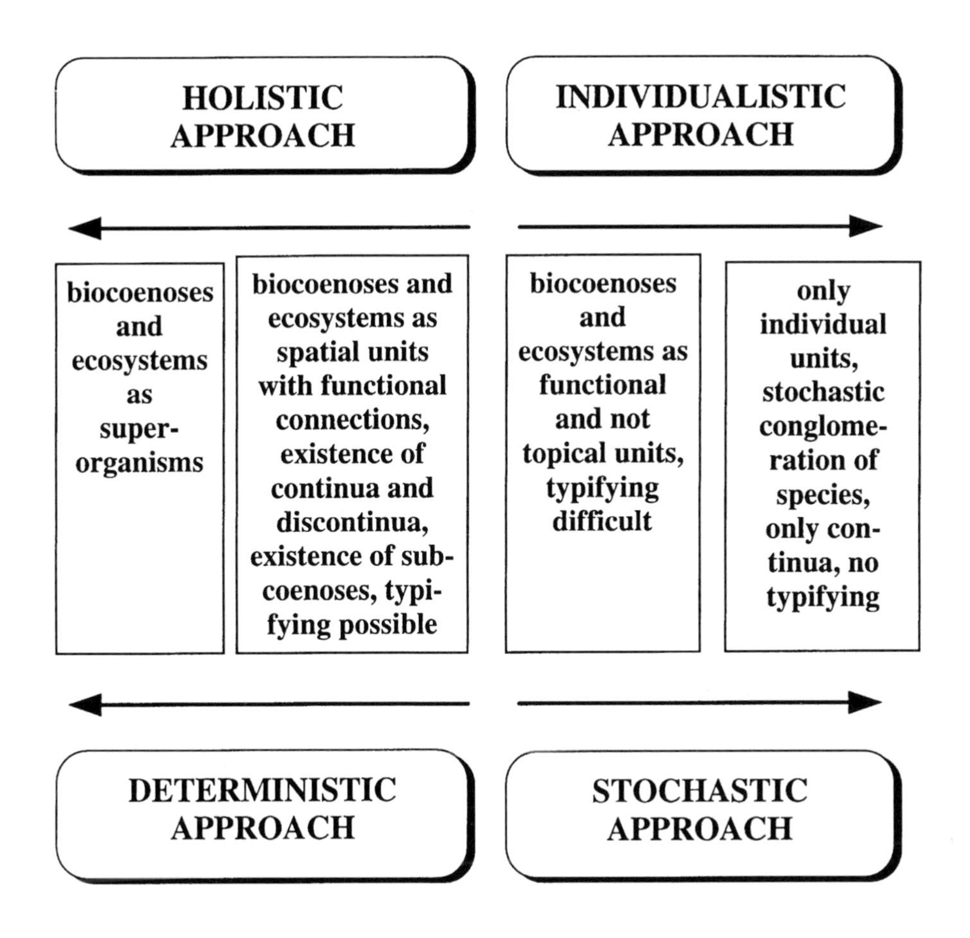

Figure 4. Differences in scientific approaches to the investigation of biocoenoeses and ecosystems: a) holistic and individualistic approach, b) deterministic and stochastic approach.

Forms of diversity

The different forms of biodiversity may basically be assigned to four types:
- diversity of elements (element pattern of biodiversity)
- diversity of interactions (dynamic pattern of biodiversity)
- mechanisms causing diversity (causing pattern of biodiversity)
- process of functioning (functional pattern of biodiversity).

Diversity of elements (element pattern of biodiversity)

Taxonomic and syntaxonomic diversity, species and coenosis diversity. Following Whittaker (1972, 1975, 1977) various species and coenosis diversity levels can be distinguished in different spatial units: α-diversity, β-diversity and γ-diversity, complemented by another category, δ–diversity (for definition see also Schwabe in this volume).

α-diversity refers to the species diversity of a certain area. It is for instance described by several calculation methods and the determination of indices (see *e.g.* Krebs 1989). One problem is how to delimit a specific area. It may be characterized according to the different spatial structure types. In this case, however, a "quasi-homogeneity" of the habitat must be presupposed, provoked by the physiognomy of a specific plant community or by certain synusia, of which this community is composed. Following v.d. Maarel (1988) α–diversity is defined as "diversity within a community"; it could also be described as "intrabiocoenotic diversity".

Gradients between different biotopes (habitats) can be analyzed by β–diversity. This procedure is especially suitable for regions with ecological gradients (ecoclines), *e.g.* forest/open land areas, zonation complexes at water banks etc., less however for areas with pronounced discontinuities. V.d. Maarel (1988) defines this diversity type as "diversity between communities", although it would certainly be more precise to describe it as "gradient diversity between communities".

γ–diversity characterizes the diversity of landscapes, in which case a landscape is defined as landscape part (= physiotope, see Schwabe in this volume). Such a landscape part consists of several communities, the entirety of which makes up a vegetation complex. In a physiotope certain uniform factor combinations can be found (geological substratum, soil conditions, nutrient balance, water balance etc.). Units relevant for the investigation of γ–diversities would be ecosystems and ecosystem complexes. Following v.d. Maarel (1988) one might speak of a "diversity of complex communities" ("interbiocoenotic diversity").

As suggested by Goetze & Schwabe (1997), γ–diversity may again be divided into γ_1-diversity and γ_2-diversity. γ_1-diversity characterizes the number of vegetation types in a vegetation complex, γ_2-diversity the number of vegetation complexes in a landscape part.

δ-diversity characterizes (analogously to β-diversity, where changes in the number of species along an ecological gradient are analyzed) changes in the number of vegetation types along an ecological gradient (Goetze & Schwabe 1997).

Sigmasociological methods are used to register and analyze vegetation complexes, and to characterize γ–diversity and δ–diversity (Schwabe 1990, 1991a, 1991b; Schwabe in this volume; Goetze & Schwabe 1997).

Diversity of life-forms. The concept "life-form" comprises the whole complex of species-specific qualities of an organism, which developed in adaptation to the particular conditions of a certain habitat (morphological, physiological, and ethological characteristics). Such life-forms can be typified. A "life-form type" belongs to a group of species, which often have different systematic ranks, but have acquired, adapting to the conditions within a habitat, analogous morphological, physiological, and ethological characteristics and modes of life in the course of evolution, and thus have the same life-form. For animals, life-form types can be classified according to feeding habit (*e.g.* phytophagous, zoophagous, parasitic, detritophagous; filter feeders, substrate eaters, grazing animals, sap feeders, stinging suckers, gatherers, predators, trappers, parasites), according to mode of locomotion (*e.g.* burrowing, crawling, climbing, jumping, flying and running animals), and according to place of residence (edaphon, atmobios, herbicolous organisms = living on or in plants; phyllobios, lignicolous organisms = living on or in wood, epizoa, endozoa, and others) (see *e.g.* Tischler 1949).

For plants, different life-forms can be distinguished according to the way of surviving the unfavourable season (classification after Raunkiaer 1907/1937), according to adaptations of the water balance (xerophilous, mesophilous, hygrophilous, hydrophilous), according to light requirement (heliophytes, skiophytes), according to soil factors, and according to diet (see *e.g.* Strasburger 1991).

A very comprehensive system of different life-form types was presented by Koepcke (1971, 1973, 1974).

Diversity of spatial structures. After Tischler (1949) a habitat can be divided into three different spatial structure types: stratotope, choriotope, and merotope. Such a differentiation is essential for the recording and analysis of synusia within a biocoenosis. A detailed characterization of the different spatial structure types will be given in the chapter "Intrabiocoenotic diversity".

Trophic diversity. Classification into producer, consumer and decomposer levels with further subtypes (see *e.g.* also Cohen 1978).

Phenological diversity. Characterization of time structures, diurnal and seasonal changes, periodic phenomena within a year (*e.g.* different flowering phenologies, see Kratochwil 1983, 1984).

Genetic and population-specific diversity. Characterization of genetic variability and of the genotype spectrum, phenomena of homo- and heterozygosis and of gene drift, mutation rate of individual populations, and others (see *e.g.* Stearns *et al.* 1990; Vida 1994; Frankel *et al.* 1995); on population-specific diversity see *e.g.* Matthies *et al.* (1995).

Biochemical diversity. Characterization of different plant ingredients (*e.g.* alkaloids), partly important as "biochemical defence" against phytophages (see *e.g.* Feeny 1976) or scents as attractant for flower-visiting animals (Kugler 1970).

Diversity of interactions (dynamic pattern of biodiversity)

Among themselves, species create bi- and polysystems and thus form so-called biocoenotic links. These interactions between the organisms induce the emergence of characteristics which may contribute to stabilizing the system (quasi-stability in the species composition). Such interaction patterns can be divided into probioses (mutualism, symbiosis, commensalism) and antibioses (predation, parasitism etc.).

Mechanisms causing diversity (causing pattern of biodiversity)

Basically two different processes causing biodiversity can be distinguished:
- effects in evolutionary times (separation, speciation, and radiation)
- effects in ecological times.

Effects in evolutionary times. In evolutionary time periods, biodiversity is attained by speciation (allopatric, sympatric). Of great importance are in this case the separation of originally linked populations, the subsequent differentiation of the separated populations, the development of isolation mechanisms, and the formation of different ecological niches. A decisive factor for high diversity rates is an only slight extinction.

An especially high species diversity is elicited by radiation. Examples for this are Darwin's Finches (Geospizinae) on the Galapagos Islands (Lack 1947), or the honey-creepers (Drepanididae) and fruit flies (Drosophilidae) of Hawaii (Mayr 1943; Carson & Kaneshiro 1976; Carson *et al.* 1970).

Effects in ecological times. In ecological time periods, a biocoenosis rich in species can only develop when communities immigrate and are newly formed. In this context, the number of ecological niches to be realized plays a decisive part. The concept "ecological niche" is used in the sense of Günther (1950). According to his definition, the ecological niche is no spatial unit, but the dynamic relation system of a species with its environment. It is composed of an autophytic/autozooic and an environmental dimension. The autophytic/autozooic dimension comprises the phylogenetically acquired morphological and physiological (for animals also ethological) characteristics of the species, the environmental dimension the sum of all effective ecological factors. Where both dimensions overlap, the ecological niche of a species is realized. The breadth of the niche depends on the degree of specialization of the ecological niches which realize it. Niche overlaps can only be tolerated by species with a greater niche breadth.

Process of functioning (functional pattern of biodiversity)

The question to what extent biodiversity contributes to the functioning of biocoenoses is controversially discussed. There is no doubt that many organism species are constantly linked by certain interactions, and that these interactions may be obligatory. Such an interaction structure has only system character when it can be differentiated from other systems and when an independent matter flow is ascertainable. The differentation of biocoenoses and ecosystems, however, has first a merely hypothetical character. Therefore only theories can be developed in reply to the questions how much redundancy a biocoenosis or an ecosystem may tolerate without being impaired in the maintenance of their functional balance, or whether there are upper and lower limits of biodiversity. The "theory of biodiversity" is closely linked with the "ecosystem theory".

The more diverse the system, the more diverse must be its functional structure to stabilize the system. The element pattern (see chapter "Diversity of elements") and the diversity of interactions (chapter "Diversity of interactions") primarily contribute to this stabilization.

Matter (nutrient) and energy flow are required to keep up the system and attain a quasi-stability. The stabilization processes include matter and nutrient absorption, transformation, and transfer (as input-output reaction).

Intrabiocoenotic diversity

A biocoenosis is composed of the plant community (phytocoenosis) colonizing a phytotope, and the animal community (zoocoenosis) inhabiting a zootope. Owing to the physiognomicly dominating higher plants, plant communities can be more easily analyzed and typified. The number of associations is remarkably high: in Germany, there are approximately 700 plant communities (Pott 1995) and more than 3,200 higher plant species (Oberdorfer 1994).

All the same, phytocoenoses can be more easily recorded in their diversity than zoocoenoses. In Germany alone, more than 45,000 animal species occur. The animal species number of a beech forest roughly corresponds to the total number of Germany's plant species. How can this wide variety of animal taxa possibly be registered?

There are different pragmatic approaches to the study of biocoenoses and their diversity:

- investigation of taxonomic groups (zootaxocoenoses): classifying biodiversity
- investigation of functional groups or guilds, respectively ("subsystems", smaller units, functional groups of co-existing species which use the same resources in a similar manner): functional biodiversity
- investigation of certain relations (*e.g.* plant-insect complexes, food chains, food webs): interaction biodiversity
- investigation of microhabitats (= synusia): classifying microhabitat biodiversity.

More than 90 % of all terrestrial animal species are bound to habitats characterized by their vegetation. The first step in the recording of an animal community must be a phytosociological characterization of the habitat, for plant communities or vegetation complexes characterized by plant communities constitute typifiable units under ecological, structural, dynamic, chorological, and syngenetic aspects (Kratochwil 1987, 1991a). Such a characterization of a habitat via its plant communities and plant community complexes is the starting-point for a registration and analysis of biocoenological (community-ecological) diversity.

The second step is a classification into microhabitats (= synusia); this classification should be based on three different spatial structure types (Tischler 1949): stratotope, choriotope, and merotope (Fig. 5).

The different strata, *e.g.* of a forest, are designated as stratotopes; here it can be distinguished between tree stratum, trunk stratum, herb stratum etc., each colonized by its own stratocoenosis. Choriotopes, on the other hand, are independent vertical structures of the entire spatial unit or of parts of the stratotope, so-called choriocoenoses, like the insect community of a tree or a shrub. Finally, in a habitat rich in structures, merotopes can be found, *i.e.* structure elements within a stratotope or a choriotope, like organisms living on leaves or on bark, or flower visitors.

STRATOTOP	STRATOCOENOSIS	EXAMPLES
horizontal structures	subcommunity belonging to the stratotope	litter layer herb layer shrub layer treetop layer

CHORIOTOPE	CHORIOCOENOSIS	EXAMPLES
typical vertical structures of the whole spatial unit or parts of the stratotope	subcommunity belonging to the choriotope	tree tree stump shrub carcase excreta anthills bird's nest

MEROTOPE	MEROCOENOSIS	EXAMPLES
structure element within a stratotope or choriotope	subcommunity belonging to the merotope	leaf residents wood residents bark residents flower residents flower visitors

Figure 5. The three different spatial structure types (stratotope, choriotope, and merotope), the coenoses they comprise (stratocoenoses, choriocoenoses, and merocoenoses), and examples for these types (based on Tischler 1949).

Stratocoenoses. Analyses of taxonomic biodiversity demonstrate that each of these strata has its own animal species inventory, *e.g.* own spider stratocoenoses in Central European oak-birch forests (Fig. 6); see Rabeler (1957). Comparisons of the strata of various plant communities, of the leaf and soil strata of a melic grass-beech forest (Melico-Fagetum), and of an oak-hornbeam forest (Querco-Carpinetum) show distinct differences in the species composition of earthworms in the stratocoenoses, especially in the leaf litter stratum (Rabeler 1960; see also Kratochwil 1987).

Choriocoenoses. Other structural elements include special, clearly differentiable elements, so-called choriotopes: a tree, a shrub, or a single plant, *e.g.*, each with its community of phytophagous insects (phytophage complex). The diversity of a choriotope will be demonstrated at the example of a bird's nest (Fig. 7); Aßmann & Kratochwil (1995); Kratochwil & Aßmann (1996a). Bird species utilize very specific requisites to build their nests. The Long-Tailed Titmouse (*Aegithalos caudatus*) builds highly characteristic nests in juniper (*Juniperus communis*) in northern Germany.

An analysis of the nesting material shows that it consists of specific materials: certain moss species, lichen species, algae etc. The composition depends on the plant community, in which the nest is built. It is an orderly, habitat-typical structural diversity. The nest of a Great Tit (*Parus major*) is built in another way, moreover this bird species is mainly found in quite different habitats. The diversity of species entails a diversity of the small structures created by them.

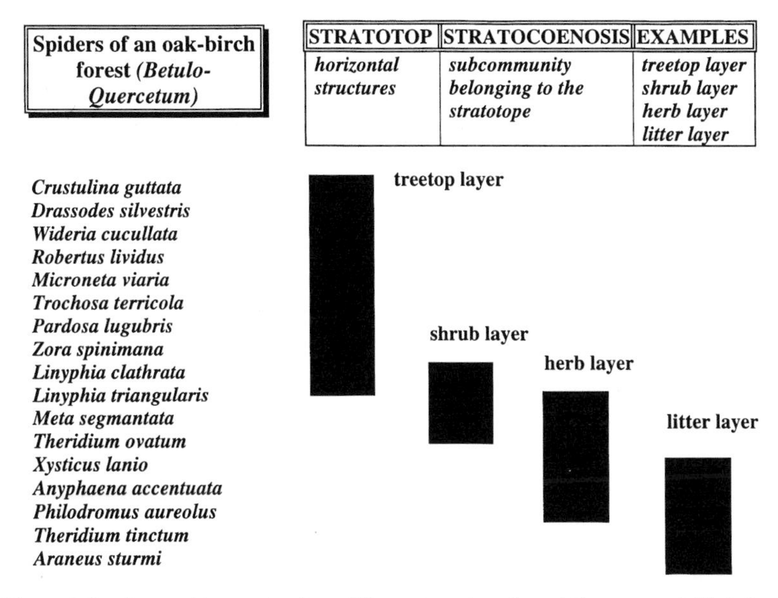

Figure 6. Dominant spider species from different strata in a Central European oak-birch forest (based on Rabeler 1957).

Merocoenoses. The merotopes, parts of strato- and choriotopes, represent the third, final element. Strato-, chorio- and merotopes combine to a special degree structural and functional diversity. Here we particularly investigate ecological niches, interaction levels, and relation structures between organisms.

The community of flower visitors corresponds to a merocoenosis, with the flowers representing merotopes. First we find a "systematic biodiversity" of very different animal groups: Hymenoptera Apoidea, Hymenoptera Aculeata, Lepidoptera, Coleoptera etc. (Kratochwil 1984). Within this flower/flower visitor system, there is a "functional diversity" introduced by the visitor: *e.g.* food relations (pollen, nectar, oil), or certain other resource relations, like the use of the flowers as warming-up places, due to their parabolic mirror-like forms (Hocking & Sharplin 1965; Kevan 1975), as "rendezvous" places (Eickwort & Ginsberg 1980), as food source for predators and parasites (Mayer & Johansen 1978; Morse 1984), as overnight accommodation *e.g.* of bees (Dafni *et al.* 1981), or as provider of nesting materials (Benno 1941). Flowers even supply scents, used to mark swarming paths, as done by the neotropic, scent-gathering euglossine bees (Euglossinae); see Evoy & Jones (1971). Alone for oil-producing plants, about 1,400 plant species (belonging to ten families) are known world-wide, and approximately 300 wild bee species specialized on them (Vogel 1988).

	Longtailed titmouse (*Aegithalos caudatus*)		Longtailed titmouse (*Aegithalos caudatus*)		Great tit (*Parus major*)
	JUNIPER PLANT COMMUNITY (*DICRANO-JUNIPERETUM*)		JUNIPER PLANT COMMUNITY (*ROSO-JUNIPERETUM*)		BEECH FOREST (*PERICLYMENO-FAGETUM*)
NESTING MATERIAL	NEST 1	NEST 2	NEST 3	NEST 4	NEST 5
BRYOPHYTES, LIVERWORTS					
Hypnum mamillatum	8,9 %	23 %	12,8 %	38 %	+
Dicranum scoparium	++	++	-	-	-
Campylopus introflexus	+	+	-	-	-
Lophocolea bidentata	+	-	-	-	-
cf. *Dicranoweisia*	+	-	-	-	-
Hypnum mamillatum	-	-	+++	-	-
Eurhynchium swartzii	-	-	++	-	-
Dicranoweisia cirrata	-	-	-	+	-
Rhytidiadelphus squarr.	-	-	-	-	50 %
Scleropodium purum	-	-	-	-	40 %
Brachythecium rutab.	-	-	-	-	+++
Eurhynchium praelong.	-	-	-	-	+
Pleurozium schreberi	-	-	-	-	+
LICHENS	11,5 %	13,1 %	14,6 %	9,8 %	-
Hypogymnia physodes	x	x	x	x	-
Parmelia sulcata	x	x	x	x	-
Physcia tenella	-	-	x	x	-
Physcia adscendens	-	-	x	x	-
ALGAE					
Ulothrix spec.	+	-	-	++	-
Pleurococcus vulgaris	+	-	-	++	-
MISCELLANEOUS					
feathers	15,8	19,4	35,5 %	11,6 %	-
juniper bark	64 %	12,6 %	9,7 %	4,7 %	-
juniper berries	++	+	+	+++	-
juniper needles	++	+	+	+++	-
spiders threads	+	+++	+++	++	-
insect cocoons	-	++	-	-	-
heather	-	+	-	-	-
indefinable	-	32 %	32 %	-	-
leaf remains	-	-	-	+	-

Figure 7. Structural diversities of nests of Long-Tailed Titmice (*Aegithalos caudatus*) and a comparison with the nest of a Great Tit (*Parus major*); based on Aßmann & Kratochwil (1995) and Kratochwil & Aßmann (1996).

Moreover also plants show different degrees of functionality (functional diversity). For the plant, the margin ranges from symbiotic relations, in which case the pollen-transferring insects are rewarded with food, to parasitism, which can be found in its most distinctive form in specimens of the genus *Ophrys*: the flowers imitate female bees and "sneak" by optical, olfactory, and tactile stimuli into the instinctive behaviour of male bees to ensure a transfer of pollinia (Kullenberg 1961; Paulus 1988).

Species diversity and functional diversity always correlate with structural diversity. One example for this is the correlation between the structure of the pollen-gathering device of a bee and certain pollen grain structures (Paulus 1978). The bee *Lasioglossum lineare* (Halictidae), main pollinator of the pasque-flower *Pulsatilla vulgaris* (Ranunculaceae), has a specific gathering device at its hind legs, composed of particularly fine hairs. These hairs exactly fit into the sutures of the pollen grains of *Pulsatilla vulgaris* (Kratochwil 1988a); a coevolution between specific flowering plant and pollinator.

How much structures determine functions, and vice versa, is shown by the next example. There are hairs on the *Ophrys* flowers. The position of a male bee landing on such a flower depends on their orientation. In this way it is also determined whether the pollinia are attached to the head or to the abdomen (Paulus 1988).

The structural diversity of a flower/flower visitor merocoenosis is immense:
- optical diversity: the colours of the flowers in the visible, but also in the ultraviolet wave range (literature in Kratochwil 1988b, 1991b)
- olfactory diversity: the multitude of different flower scents (Kugler 1970)
- ethological diversity: the behavioural variety of flower visitors (Westrich 1989)
- phenological diversity: the diurnal and seasonal variation of the occurrence of flowers and their pollinators (see *e.g.* Kratochwil 1983).

Each plant community has its own animal community or synusia of different animal communities. On ecosystem level, the structural and functional diversity levels of different organism groups correlate with their specific abiotic environment. The biocoenoses or biocoenosis complexes are characterized by certain character species. However, each biocoenosis has its own range of diversity types and patterns. The greater the species diversity, the more varied are other diversity types: genetic diversity, space-structural and physiognomic diversity, biochemical diversity, phenological diversity etc.

Interbiocoenotic diversity

As a rule, landscapes are not composed of single biocoenoses, but of biocoenosis complexes and a mosaic of different ecosystems. The development *e.g.* of individual vegetation units into associations is not arbitrary, but follows certain rules (for a detailed depiction see Schwabe 1990, 1991a, 1991b; Schwabe in this volume). Especially ecosystems with a distinct microgeomorphology, *e.g.* inner-alpine dry slopes (Schwabe 1995; Schwabe *et al.* 1992) or steppe heaths in Central Europe (Köppler 1995; Köppler & Schwabe 1996) are perfect examples of habitats with very high species and coenosis diversities in Central Europe.

It is interesting that regularities on the species/biocoenoses and biocoenoses/ biocoenosis complex levels follow the same natural laws (Schwabe & Kratochwil 1994). At the example of the rock and moraine physiotopes in central alpine dry areas it could be shown that "Thienemann's 2nd basic principle" may also be formulated as coenological and landscape-ecological principle: "The more variable the environmental conditions of a habitat complex, the larger the number of its coenoses/synusia." and: "The more the environmental conditions of a habitat complex deviate from the normal and for most coenoses/synusia optimum conditions, the poorer in coenoses/ synusia the complex will become, the larger and the more characteristic are the occurring coenoses/synusia."

On the other hand it is just these areas characterized by vegetation complexes which represent habitats for certain animal species (Schwabe & Mann 1990; Schwabe *et al.* 1992) and for special zoocoenoses (Kratochwil 1984, 1989; Kratochwil & Klatt 1989a, 1989b; Aßmann & Kratochwil 1995; Kratochwil & Aßmann 1996a, 1996b), and thus allow a biocoenological analysis of landscape units.

Now ecosystems are not static structures, but dynamic units, be it in the scope of succession or of spatial and temporal cyclic processes. Particularly in Central and southern Europe, man has played an important part as "landscape architect", for he created, by extensive and long-lasting agricultural management, a great biodiversity. This positive influence was diminishing in the course of the past century, when intensive management forms were introduced and mechanization set in.

A good example for habitats created by man are the pasture-woodlands, *e.g.* in northern Germany (Aßmann & Kratochwil 1995; Kratochwil & Aßmann 1996a, 1996b; Pott in this volume). Despite intensive interventions of agriculture and forestry, primarily in the past fifty years, there are still some habitats in the north-west German lowlands which reflect in a special way the long-lasting and extensive anthropogenic utilization of the land: the "pasture-woodlands". They arose due to range management (pasture farming) that was first restricted to woodland sites (wood-pasture), but then increasingly led to an opening of the woodland and to the development of numerous open land sites, also as a consequence of further utilizations (*e.g.* cultivation by sod). What is particularly striking for visitors still today is the impression of a "parkland", a mosaic of sand dunes free of or poor in vegetation, extensively managed open pastures with grass, tall herb communities ("hem communities"), richly structured edges of woodlands, the occasional individual tree, clusters of shrubs and trees of varying sizes and woodland communities with open stands, and a large share of especially characteristic tree individuals, which often still show signs of the former wood-pasture: pollard forms, trimming and pruning marks, as well as distinct traces of the damage caused by browsing animals (see *e.g.* Burrichter 1984).

The high biological diversity of such a landscape is due to its richness in structures, both on the species (α–diversity) and on the coenosis level (γ–diversity), and to its gradients (β–diversity, δ–diversity). These structures arose as a consequence of anthropo-zoogenic landscape dynamics. The landscape genesis follows the principle of "variety in space" (a high degree of constant spatial changes of the factor combinations, van Leeuwen 1966) and is characterized by a continuous preservation of its mosaic structure. The stabilizing and system-preserving factors are in this case not the natural factors, as asserted by Remmert (1991) for woodland ecosystems, but the anthropo-zoogenic influences which have affected the biocoenoses since the Neolithic Period. Judging by pollen-analytical findings, the areas were being used as pasture-woodlands for about 5,000 years (Pott & Hüppe 1991).

Hypotheses on diversity

There are a number of different hypotheses on biodiversity (see *e.g.* Solbrig 1991, 1994; Schaefer in this volume, and others). They constitute the basis for the development of a general "theory of biodiversity". In the following 30 hypotheses will be presented (formulated as questions), some of which correlate as to their contents. Except for the first one, all hypotheses refer to the ecosystem and the ecosystem complex level. Their order is arbitrary and does not reflect an evaluation. It should also be stressed that none of the hypotheses is absolutely valid; they apply as a rule, but allow for exceptions.

1 *Is the entire biodiversity (holodiversity) increasing with higher levels of hierarchy in a system?*

If a systemary approach is taken as basis and the respective elements of a hierarchy level can be integrated into the next higher level (see Fig. 3), the degree of complexity is increasing on higher levels of hierarchy, since on every new level further new system characteristics (emergent characteristics) appear ("principle of functional integration" after Odum 1971). The entirety of the elements of one level is more than merely the sum of its components. However, in the framework of our examination, this principle only applies to the level of biodiversity. The biochemical, genetic, or structural diversity of a landscape part is always much greater than that of an individual organism or of a cell. The principle is moreover not wholly applicable to compensation phenomena. The physiological constitution of a single organism, for instance, is as a rule always greater than its ecological constitution (restriction of the physiological constitution, *e.g.* under prevailing competitive conditions).

2 *Is the species and ecosystem diversity increasing with advancing age of the ecosystem?*

Ecosystems like the tropical rain forest, the coral-reefs, old lakes (*e.g.* Lake Tanganyika, Lake Baikal) have, owing to their advanced age, created ecosystem complexes with a particularly high diversity. According to this "time hypothesis" (Latham & Ricklefs 1993) older ecosystems have more species than younger ones. This is on the one hand due to evolutive reasons (speciation), on the other to ecological reasons (immigration and colonization). Prerequisites for an augmentation of the species number are that, with growing succession of a habitat and its microhabitats, its habitat diversity increases, that a species pool exists, from which species can immigrate and realize ecological niches, and that the new "inhabitants" are compatible with the others and may be incorporated into their interaction structure (Cornell 1993).

With advancing age of an ecosystem, the share of organisms following a K strategy is increasing, the share of the so-called r strategists is decreasing (see hypotheses 14 and 17). It is undisputed that a certain "species set" is essential for the maintenance of the homeostasis of an ecosystem, but also for the genesis of a biocoenosis. Whether the entire species diversity of such systems necessarily has a system-preserving character, is controversially discussed. It may however be assumed that *e.g.*

a historically caused high diversity often has redundancy character (see hypothesis 16). Individual relic biocoenoses may be very rich in species, too (see hypothesis 12).

However, there are also examples to the contrary, showing that in the course of succession biodiversity must not necessarily increase towards a climax stage ("intermediate disturbance" hypothesis) (Connell & Slatyer 1977; Connell 1978; Huston 1985); see hypothesis 27. So Pignatti & Pignatti (in this volume) could demonstrate for Mediterranean ecosystems that the species diversity is not necessarily increasing in the course of succession, that the man-made habitats are richer in plant species than the natural vegetation without any anthropogenic influence. The same may apply to individual zootaxocoenoses (Kratochwil & Klatt 1989a, 1989b).

In a beech forest, the diversity of plant species is also decreasing with succession; this is however not true for its fauna.

3 *Is biodiversity increasing with the degree of biocoenotic progression?*

Analogous to the degree of sociological progression of a plant community or of a vegetation complex in phytosociology (Dierschke 1994), there are also different degrees of biocoenotic or ecosystemary progression between different biocoenoses. The concept "degree of sociological progression" involves:

- attachment of the majority of the individuals from one community, of stands and communities, to a certain site
- interrelations between the individuals of different species and communities
- diversity of the structure of strata, diversity of life-forms
- longevity of the stands.

An augmentation of the degree of biocoenotic progression should entail an increase in biodiversity.

4 *Is ecosystem diversity increasing with growing radiation energy and humidity?*

Although very little energy is needed for the photosynthesis rate of autotrophic organisms (often less than 1 % of the global radiation), the "operation temperature" must be favourable for the constructive and the energy metabolism throughout the year, to provide advantageous environmental conditions, especially for ectothermic organisms. That is why it is not surprising that particularly the tropics, as regions with a diurnal and not a seasonal climate, with high temperatures and a high amount of precipitation, are the centres of greatest biodiversity on earth. This also reflects thermodynamic natural laws. A very high degree of systemary order (neg entropy) presupposes the supply of a large quantity of free enthalpy. Accordingly Pignatti & Pignatti (in this volume) could, when assessing the biodiversity of Mediterranean ecosystems, prove that there is a relation between the number of plant species and temperature and water availability. The amount of precipitation alone is not significant, the rate of evaporation is the decisive factor (Pignatti & Pignatti in this volume). Thus the overall productivity of an ecosystem is causally related to its richness in species (Currie 1991; Latham & Ricklefs 1993). The degree of quantity and quality of primary production determines biodiversity.

5 *Is ecosystem diversity increasing with relief intensity?*

An increase in biodiversity may be related to a higher relief intensity, since in the temperate zones the angle of incident radiation is getting more favourable with higher inclination (see also hypothesis 4). In the northern hemisphere, southern slopes are richer in species than northern ones. The same is true for Mediterranean ecosystems, as shown by Pignatti & Pignatti (in this volume). With varying relief intensity, different meso- and microclimate conditions, as well as different soil conditions (soil types) alternate on a small scale; thus different site conditions and subsequently a differentiated vegetation develop. The water factor has a modifying effect.

6 *Is ecosystem diversity dependent on the geological set-up and its diversity?*

The distribution of numerous plant species depends on certain geological and pedological conditions. Consequently they are, according to their respective, geologically determined life-form type, designated as calcicolous plants, silicate plants, chalkophytes etc. In an area where geological and pedological patterns intensely change, more different plant species grow than in a region with a smaller diversity of geological and pedological site factors.

As many molluscs and arthropods need $CaCO_3$, they are richer in species in calcareous than in siliceous regions.

In connection with the different geological, but also pedological characteristics of an area, the respective pH-value determines its diversity, too. As a rule, higher species numbers are attained in a neutral or slightly alkaline milieu than in an acid environment or one extremely rich in bases.

7 *Are species and ecosystem diversity increasing with the possibility of a post-glacial recolonization?*

Many areas in the northern temperate zones could, owing to a glaciation in the Pleistocene, not be colonized by plant and animal species. The ice ages moreover led to the local extinction of numerous species. This is especially true for Europe, where the Alps - as a kind of crossbar - prevented the escape of these species to more southern regions (refuge areas). In northern America, where the great mountain ranges run from north to south, this was not the case. That is why - so many scientists assume - the diversity of a number of tree species is higher in northern America (Walter & Straka 1970). In addition, the conditions for a recolonization were more favourable in northern America than in Central Europe.

8 *Is diversity increasing with the probability of allopolyploid and autopolyploid processes?*

The fact that in the course of the ice ages numerous populations were divided into subpopulations led to gene drift and other processes (different selection pressures, random selection) and, in consequence, to a greater variation of genotypes. In extreme cases, the formation of new species was only made possible by allo- and autopolyploid processes (species and subspecies level); see *e.g.* Ehrendorfer (1962). Man has also considerably influenced this development (Pignatti 1983).

9 *Is biodiversity increasing with the size of an area?*

Inspired by the studies of Arrhenius (1921) and Palmgreen (1925) of terrestrial plants, species-area relations have long since been described for highly different organism groups from numerous islands in the sea, but also from continental islands of varying sizes. So Darlington (1957) could show that on the West Indies the species number of reptiles and amphibians is rising with increasing island size; similar phenomena were observed for the birds on the Solomon Islands (Diamond & Mayr 1976) or for the higher plant species of the Azores (Eriksson *et al.* 1974). A number of further examples substantiating this thesis can be given (Diamond 1972; Lassen 1975; Galli *et al.* 1976; Aho 1978; Jurvik & Austring 1979 etc.).

This species-area relation may be expressed by the simple formula $S = C \cdot A^z$, with S being the species number, A the size of the island, and C a constant which depends on the respective biogeographical region of the investigated taxon. Another parameter, the exponent z, ranges, according to empirically gathered data, as a rule between 0.20 and 0.35, independent of the studied taxocoenosis, be it ground beetles, ants, birds, mammals, or plants (Connor & McCoy 1979). The exponent indicates the inclination of the regression line log S to log A (Preston 1962), which is also influenced by the constant C. Within a terrestrial area, the species number is also rising with increasing area size, however, the z-value is far lower in this case. This is *e.g.* proven by a comparison of the relation of area and number of ant species (Ponerinae and Cerapachynae) from different-sized Molucca and Melanesian islands with the species-area curve under non-separated conditions, *e.g.* on New Guinea (Wilson 1961). The z-values of the species-area curves on the continent amount to merely 0.12-0.17; see also compilations of the z-values of different taxonomic groups in MacArthur & Wilson (1967) and May (1975), quoting original literature.

Preston (1962) could mathematically derive the z-value, on the hypothetical assumption that both the species and the individual numbers are lognormally distributed in a site. Studies of birds by Preston (1962) and of moths by Williams (1953), as well as of many other animal groups, support this assumption, but there are also exceptions. Preston (1962) calculated, via a canonical distribution, a z-value of 0.263 for insular relations, which is in good agreement with many values ascertained in nature. All these examples prove that biodiversity is increasing with growing size of an area (but see also Haeupler 1997).

10 *Is habitat diversity increasing with the size of an area?*

The relation between species number, growing size of an area, and increase in habitat diversity has often been discussed in the literature, however, there are considerable differences in the assessment of the importance and weight of the main factors. Some authors consider area size and habitat diversity to be exchangeable (Hamilton *et al.* 1964; Johnson & Raven 1973; Simberloff 1974). Others see them as extremely correlating factors, with one giving rise to the other. In this case, there are different views as to their importance: Johnson & Simberloff (1974), Simberloff (1974), Reed (1981), Lynch & Whigham (1984) consider the habitat diversity to be of greater relevance, whereas Hamilton *et al.* (1964), Johnson & Raven (1973), Brown (1971) argue that the area size is more important.

Based on the ideas of Dean & Connell (1987a, 1987b), O'Connor (1991) developed two alternative hypotheses which may explain the relation between species diversity and area size. The "sampling phenomenon hypothesis" purports that the relation between the increase in habitat diversity and the increase in species diversity is alone determined by the size of the investigated area and the consequently higher number of available resources (Douglas & Lake 1994). The "resource availability hypothesis" argues that the new resource qualities bring about the increase ("niche availability hypothesis" in the sense of Dean & Connell 1987a, 1987b). Buckley (1982, 1985) also points out that not the diversity of the whole area is decisive, but the quality of single habitat types.

The species number (plant and animal species) of an area with homogeneous environmental conditions is not or only slightly increasing, even if a considerably larger area of the same quality is investigated (Vestal & Heermans 1945; Vestal 1949; Goodall 1952; Greig-Smith 1964; Forman *et al.* 1976; Dierschke 1994). Therefore it is essential to know the minimum area of a plant community or biocoenosis, beyond whose limits the number of typical, characterizing species is no longer rising. The same applies to communities which form typifiable vegetation complexes: their number varies merely slightly in a specific landscape (Schwabe & Kratochwil 1994).

11 *Have separated ecosystems lower species numbers than less separated ones?*

At the same area size, less separated islands have higher species numbers than those which are far away from a colonization source (the same is partly true for continental islands). This was *e.g.* shown by Lack (1969) for the avifaunas of different islands off New Guinea. Islands lying more than 3,200 km away from the mainland had a much lower species number than those less than 800 km away. A different "reachability" of islands for immigrating ground beetles (Carabidae) was already shown by Lindroth (1960); this has meanwhile been described in many cases and for many taxa. Rosenzweig (1995) however points out that not the distance from the mainland can be regarded as a generally comparable measure for the respective species number of an island in the sea, but the different immigration probabilities. They are indirectly linked to the distance, however, they depend on the "quality" of the "source".

12 *May the species oversaturation of an area be one reason for an especially high species number?*

In accordance with the island theory (equilibrium theory) by MacArthur & Wilson (1967), many cases are meanwhile known in which, due to an increase in the sea-level, land bridges sank in the course of the past 10,000 years, or islands were reduced in their size. The biocoenoses there are at present still species-oversaturated, since extinction exceeds colonization and an equilibrium has not yet been reached. Diamond (1972, 1973) studied 32 of such former "land bridge islands" off the shore of New Guinea, which were linked to the mainland only 10,000 years ago. The changes in the sea-level in the continental shelf zone are quite well documented; considering several analyses made by different authors, it varied by at least 60 m in the last 10,000 years. An avifaunistic investigation performed by Diamond (l.c.) only showed a relation between species number and area size; the z-value amounts to more than 0.35. By increase in the sea-level, destruction of the former land bridge and an entailing reduction in size of the area, these islands which previously belonged to the mainland are for the moment species-oversaturated. Owing to the now lower area size and in accordance with the species-area relation, a lower species diversity is to be expected. Until a new equilibrium is attained, extinction prevails; a further colonization is restricted by the barrier effect. Terborgh (1974) who studied the avifauna of five neotropic land bridge islands obtained the same results (see also Karr 1981).

This phenomenon was also observed in some relic biocoenoses. From the Great Basin in western USA, 17 mountain ranges rise with heights of over 3,000 m. The boreo-alpine habitats on the summits of these mountain islands are today surrounded by dry and hot sites. A link to the extensive Rocky Mountains and the Sierra Nevada was only present in the Pleistocene. The reduction in size of the area, and the lack of colonization possibilities and colonization ability of the species led to a high number of small mammals (after Brown 1971).

13 *Is species diversity increasing with habitat and structure diversity?*

According to the "habitat diversity hypothesis", set up by Lack (1969), species diversity is increasing with habitat and structure diversity (see in this context also Hamilton *et al.* 1964; Simberloff 1974, 1976; Tangney *et al.* 1990; Hart & Horwitz 1991; Kohn & Walsh 1994). This hypothesis correlates with the "niche theory" (see hypothesis 14).

14 *Is the number of ecological niches related to the number and composition of the species present in a biocoenosis?*

A high diversity of ecological niches may be due to a (historical) competitive situation. After niche differentiation, competition among the species is reduced, and a coexistence of different species is possible. This however only happens in localities where the immigration of species plays no dominant part in establishing a species community. Processes of niche differentiation take a longer time.

A high species number in a habitat is as a rule based on a niche differentiation (niche partitioning, see Schoener 1986). The degree of the ecological occupation of a niche always depends on specialization and a reduction of the competitive pressure between the species. According to the "competitive exclusion principle", an increase in diversity should be accompanied by a decrease in interspecific competition. In a "mature" ecosystem, the share of stenoecious species should be higher than that of euryoecious ones. Likewise a lower number of individuals of numerous stenoecious species correlates with a higher number of individuals of few euryoecious species (see Thienemann 1920, 1956); see also hypothesis 15. At an early succession stage, however, euryoecious species prevail as a rule, consequently the probability of a first colonization by such r strategists is higher.

The diversity of niches in a biocoenosis is determined by the following factor groups (Diamond 1988):
- quantity of the available resources and requisites
- quality of the available resources and requisites
- interactions between the species
- dynamics of the biocoenosis

Species diversity always correlates with resource and requisite diversity.

15 Is the number of individuals of certain species decreasing with growing species diversity?

If it is assumed that the carrying capacity (total number of individuals) of a habitat is limited, only a lower number of individuals of single species can occur when the species number is increasing. The respective population size is to a certain extent species-specific ("minimal viable population size").

This hypothesis reflects Thienemann's basic principles. In his summarizing work (1956) he formulated (p. 44): "There are sites of optimum favourable development for organisms... Here the conditions for life are stable, harmonic, no excess to any side; thus life possibilities for many species. But when a vital factor occurs in a lower amount or with less intensity, or when another one gains a superior position; when the optimum is shifting, ..., via a 'pejus', towards a 'pessimum': then the species number of the biocoenosis is more and more decreasing, and finally only few species remain. These may however, provided that the conditions for life are favourable, develop in enormous individual numbers, since they have no food competitors..."

Thus two biocoenotic basic principles apply (Thienemann 1920): 1): "The more variable the conditions for life of a site, the larger the species number of the respective community." 2): "The more the conditions for life of a biotope deviate from the normal and for most organisms optimum conditions, the poorer in species the biocoenosis will become, the more characteristic it gets, the more individuals of the single species will occur."

16 *Do redundancy phenomena occur more frequently with growing species diversity?*

According to the redundancy hypothesis also such species exist in an ecosystem that are not directly important for the maintenance of the system and do not influence the species structure to a great extent. The number of these redundant species should increase with growing biodiversity (on redundancy see Lawton & Brown 1993; Walker 1992).

17 *Is with increasing biodiversity the share of r strategists decreasing, while analogously that of K strategists is growing?*

With increasing niche differentiation, the share of K strategists is growing, that of r strategists decreasing. This hypothesis coincides with hypothesis 2 and hypothesis 14.

18 *Is the extinction probability of individual species growing with increasing degree of ecological niche differentiation?*

The increasing specialization of a species may in particular cases become a selection disadvantage. This can *e.g.* be shown at the phenomenon of the so-called "taxon cycle", in which colonizers on islands (as a rule r strategists and wide-spread, ecologically not differentiated species) develop into geographical subspecies, then more and more differentiate and specialize (are K-selected), and thus provoke an evolution which may lead to an endemism (Wilson 1961; Ricklefs & Cox 1972). Finally the highly specialized forms are extinguished by the competitive pressure of newly colonizing species (generally r-selected). With a new colonizer this cycle starts afresh. Examples for a "taxon cycle" can be found for birds (West Indies) (Ricklefs & Cox 1972) and ants (Melanesia) (Wilson 1961). The theory of the "taxon cycle" demonstrates how dangerous wide-spread and highly competitive "generalists" can be when they, after having overcome special barriers, "attack" extremely evolutionized systems. Simberloff & Cox (1987) and Simberloff *et al.* (1992) cite, among others, the following disadvantages of corridors facilitating the access to habitats rich in species: dispersal of pests and diseases, immigration of strong competitors, immigration of predators. A separation of single habitats is often an important protective mechanism to maintain a higher species diversity.

19 *Is the trophic structural diversity within an ecosystem (phytophages, carnivores, parasites, hyperparasites, parasitoids etc.) growing with increasing species richness?*

By trophic structural diversity, the diversity of different trophic levels is understood. Trophic diversity involves phytophages (feeders on living plant material), saprophages (utilizers of dead organic matter), microphytophages (feeders on bacteria, fungal hyphae and/or algae), and zoophages (predators, parasites and parasitoids). Especially the category of zoophages is further differentiated (zoophages of first, second, third and higher orders, parasites and hyperparasites etc.). The greater the species richness, the higher the trophic structural diversity of an ecosystem.

20 *Is biodiversity increasing in the course of the food chain?*

Within an ecosystem, the flow of energy and matter can only occur via different trophic levels. The distribution of biomass is larger for producers than for consumers, and it is further decreasing with each higher consumer level. For the species diversity, however, the reverse is true. The highest species numbers are attained by parasites and parasitoids (see also Schaefer in this volume). Zoophages are richer in species than phytophages, saprophages or microphytophages. The diversity of parasites by far exceeds that of predators. The extent of diversity on the lowest level (consumers of first order) positively correlates with the one on higher levels (consumers of higher order, parasites, hyperparasites).

21 *Is biodiversity increasing with the species richness of the respective immigration pool?*

As biocoenoses - unless they are very old, in which case evolutive reasons may be given for their original biodiversity - develop as a consequence of the immigration of species that colonize a habitat by realizing ecological niches, their composition depends on the potential immigration pool of the environment. Positive correlations between local and regional species pool could be demonstrated by numerous authors (see Eriksson 1993; Ricklefs 1987; Lawton 1990; Rosenzweig 1995). A prerequisite for a colonization is the existence of "open niches"; see hypotheses 13 and 14.

22 *Of what importance is the separation of single geographical areas for the diversity of convergent developments?*

The distribution of plant and animal species is restricted to certain geographical areas. Independent of their natural relationships, organisms may show, because of a similar mode of life and in adaptation to a similar habitat, many identical features in the form and morphology of their bodies (convergence). In the respective ecosystems, they have a similar "ecological and functional rank". The biocoenoses to which they belong are therefore also called isocoenoses. The diversity of convergent developments is due to similar ecological selection pressures on different species sets.

23 *Does an increase in species diversity correlate with an increase in the variability of the micro- and the mesoclimate?*

An increase in the structural diversity of a habitat entails a diversity of microcoenoses (synusia). Their existence is frequently due to a small-scale alternation of meso- and microclimate conditions. Different microclimate phenomena are also dependent on the soil substrate, especially in the temperate zones. Needle ice, *e.g.*, leads to a loosening of the upper soil material and to the formation of synusia of annual plant species (therophyte communities).

24 *What effect does an increase in the extensive human influence have on biodiversity?*

Pignatti & Pignatti (in this volume) could show that because of the human impact the biodiversity of the Mediterranean vegetation is far greater as under natural conditions (without anthropogenic influence), contrary to tropical regions, where highest species diversity is only attained at the climax stage. Thus a general connection between high biodiversity (α- and γ-diversity) and primary vegetation cannot be made. The increase in biodiversity by extensive human influence is due to the fact that, over different stages of succession, a climax vegetation may possess a multitude of different man-made plant communities (see also hypothesis 27: "intermediate disturbance" hypothesis; Connell & Slatyer 1977; Connell 1978; Huston 1985). The manifold anthropo-zoogenic influences (kind, point in time and extent of soil cultivation, mowing, grazing etc.) are reflected by the wide variety of vegetation selected due to these measures. The impact has a multiplicative effect and induces the origin of a number of possible vegetation types (progressive development). One important prerequisite is however that the factors do not become extreme factors with a levelling effect (recessive development); within the past fifty years, this has frequently led to a significant decrease in species in landscapes intensively used by agriculture and forestry.

25 *Is the share of species of smaller body size increasing with growing species number?*

As a rule, species of small body size dominate in a habitat (May 1978; Rosenzweig 1995). This is partly due to the fact that there is a much greater habitat diversity for small species than for bigger ones, and that thus smaller species have much better niche differentiation possibilities in the spatial and temporal axis. This ratio of small to big species is especially characteristic for sites with a great spatial heterogeneity. In addition, smaller species often have a higher mobility and thus a better immigration potential. Schaefer (in this volume) has formulated the following hypothesis: Diversity is higher for taxa containing more mobile species (see also Blackburn & Gaston 1994). Investigations of carabids (Carabidae) showed that small macropterous species are first colonizers; the share of brachypterous species, as well as that of bigger ones, is only increasing in the course of time.

Among the arthropods, particularly the relic communities have a higher share of bigger species. K strategists are usually bigger than r strategists. There are more brachypterous forms in relic communities.

26 *Is the share of species with a shorter life cycle increasing with growing species number?*

A habitat is usually dominated by species of small body size (see hypothesis 25) and short life cycles (May 1978; Rosenzweig 1995). The small body size often correlates with a short generation time (see Schaefer in this volume). This principle, which was originally applied to animal species, is also true for plant species. In a plant community, *e.g.*, the duration of the flowering stages correlates with the species richness (see also Kratochwil 1984). A niche differentiation in the time axis is thus better possible.

Spatial heterogeneity is generally favourable for organisms of small body size, however, it does not imply that all these organisms are also short-lived. If the habitat quality remains stable in the time axis, longevity or - with insects - polyvoltinism may occur. For insects, however, a high species diversity within a site can rather be attained by monovoltinism than by polyvoltinism.

27 *Do extensive disturbances by spatial and temporal heterogeneities increase the species number?*

The phenomenon that extensive disturbances by spatial and temporal heterogeneities increase the species number has been formulated as "intermediate disturbance" hypothesis (Connell & Slatyer 1977; Connell 1978; Huston 1985). A similar connection is made by the principle of "variety in time and space" (van Leeuwen 1966; see in this context also Pickett & White 1985).

28 *Are degree of disturbance and body size linked to species diversity?*

An increase in the extensive disturbance (see hypothesis 26) favours smaller species and such with shorter life cycles. This also applies to anthropo-zoogenically extensively influenced sites. In stable habitats, on the other hand, long-lived (example: lichens) and big species (example: many tree species) prevail.

29 *Do life-form diversity and species diversity correlate?*

As shown in the chapter "Diversity forms", an increasing species diversity must be accompanied by a greater variation of life-form types, since these are essentially different as to their diet and mobility. This hypothesis correlates with the hypotheses 19 and 20.

30 *Is there a correlation between plant species diversity and animal species diversity?*

As a rule there is a positive correlation between plant diversity and variety in animal species (Andon 1991; Gaston 1992). This relation is basically due to the close linkage of a number of phytophages to certain plant taxa and to predators and parasites, which occupy higher trophic levels. However, a plant stand poor in species (*e.g.* reeds with *Phragmites communis*) may cause a great animal species diversity. This can mainly be put down to the high structural diversity of the key species *Phragmites communis,* but also to environmental factors changing clinally and on a small scale (open/light and dense stands, young and old reeds, a changing reed structure [in dependence on the depth of the water, on abiotic factors, like wave action, and on biotic factors, like the influence of bird species, mammals etc.]). In beech forests, the great variety of animal species does not correlate with a high plant diversity, either.

Applied aspects of biodiversity

At present, about 1.5 million of the earth's animal and plant species have been described (Wilson 1989; Heywood & Watson 1995). Their actual number maybe varies between 5 and 30 million (see in this context also May 1988). For the last quarter of this century, scientists have predicted the extinction of approximately 1 million species

(Myers 1985). Also here an exponential tendency can be ascertained: from 1600 to 1900, every four years a species was eradicated by man, after 1900 every year; currently more than one species disappear per day. Wilson (1989) has assumed that every hour one species is extinguished; today already up to three species. According to Lugo (1992), 20-50 % of all species will have disappeared by the end of this century.

Under natural conditions, the net growth rate of the species number is 0.37 % in 1 million years, that is to say 0.00000037 %; an extremely low value. The natural extinction rate has thus been increased by 10,000-fold by man; the decrease is at least 100 times higher than the loss of species in the past 65 million years (Wissenschaftlicher Beirat der Bundesregierung "Globale Umweltveränderungen" [1] 1993; see in this context also Smith *et al.* 1993), and the rate of loss of genetic diversity on the level of populations yet extends this value by far.

The centres of especially high biodiversity lie in the tropics, mainly in the tropical mountainous areas. On few hectares of forest in south-eastern Asia or in the Amazon region, more tree species can be found than in the whole of Europe. In Venezuela's "evergreen rain forest" there are at least 90 tree species per hectare (Walter & Breckle 1984). In special regions, the loss of biodiversity is significant: world-wide, numerous ecosystem types are particularly endangered, among them the "tropical rain forests", certain marine ecosystems, islands in the sea, high mountain ranges, arctic and subarctic habitats, savannahs, steppes, and semideserts, large river systems, mangrove forests, and many lakes, but also the landscapes in the countries we live in.

A loss of biodiversity cannot be tolerated for ecological, ethical, religious, aesthetic, and cultural reasons, all the more as the destruction of biodiversity is irreversible (Arrow & Fisher 1974; Bishop 1978). To maintain biodiversity, to work out theoretical principles and translate them into practical measures is one of the major tasks of the next years. The maintenance of biodiversity is closely linked to the survival of man on earth, and has thus been incorporated into the concept of a "sustainable development".

References

Aho, J. 1978. Freshwater snail populations and equilibrium theory of island biogeography. I. A case study in southern Finland. Ann. Zool. Fenn. 15: 146-154.

Akeroyd, J. 1996. What really is biodiversity? Plant Talk 4: 2.

Andon, D.A. 1991. Vegetational diversity and arthropod population response. Ann. Rev. Entomol. 36: 561-568.

Arrhenius, O. 1921. Species and area. J. Ecol. 9: 95-99.

Arrow, K.J. & Fisher, A.C. 1974. Environmental preservation, uncertainty and irreversibility. Quart. J. Econom. 88 (2): 312-319.

Aßmann, T. & Kratochwil, A. 1995. Biozönotische Untersuchungen in Hudelandschaften Nordwestdeutschlands - Grundlagen und erste Ergebnisse. Osnabrücker Naturwiss. Mitt. 20/21: 275-337.

Benno, P. 1941. Een tweetal zeldsame bijen uit de Lyers, met aantekeningen bij een nest van *Osmia papaveris* Latr. Entomol. Ber. 238: 311-315.

Bisby, F.A. 1995. Characterization of biodiversity. pp. 21-57. In: Heywood, V.H. & Watson, R.T. (eds), Global Biodiversity Assessment.

Bishop, R.C. 1978. Endangered species and uncertainty: the economics of a safe minimum standard. Amer. J. Agricult. Econom. 60 (1): 10-18.

[1] *Scientific Advisory Board of the Government of the Federal Republic of Germany "Global Environmental Changes"*

Blab, J., Klein, M. & Ssymank, A. 1995. Biodiversität und ihre Bedeutung in der Naturschutzarbeit. Natur u. Landsch. 70 (1): 11-18.
Blackburn, T.M. & Gaston, K.J. 1994. Animal body size distributions: patterns, mechanisms and implications. TREE 9: 471-474.
Brown, J.H. 1971. Mammals on mountaintops: nonequilibrium insular biogeography. Amer. Nat. 105: 467-478.
Buckley, R.C. 1982. The habitat-unit model of island biogeography. J. Biogeogr. 9: 339-344.
Buckley, R.C. 1985. Distinguishing the effects of area and habitat type on island plant species richness by separating floristic elements and substrate types and controlling for island isolation. J. Biogeogr. 12: 527-535.
Burrichter, E. 1984. Baumformen und Relikte ehemaliger Extensivwirtschaft in Nordwestdeutschland. Drosera 1: 1-18.
Carson, H.L. & Kaneshiro, K.Y. 1976. *Drosophila* of Hawaii: systematics and ecological genetics. Ann. Rev. Ecol. Syst. 7: 311-345.
Carson, H.L., Hardy D.E., Spieth, H.T. & Stone, W.S. 1970. The evolutionary biology of the Hawaiian Drosophilidae. pp. 437-543. In: Hecht, M.K. & Steere, W.C. (eds), Essays in Evolution and Genetics in Honor of Theodosius Dobzhansky. Appleton-Century Crofts.
Cohen, J.E. 1978. Foods Webs and Niche Space. Princeton Univ. Press, Princeton.
Connell, J.H. 1978. Diversity in tropical rain forests and coral reefs. Science 199: 1302-1309.
Connell, J.H. & Slatyer, R.O. 1977. Mechanisms of succession in natural communities and their role in community stability and organization. Amer. Nat. 111: 1119-1144.
Connor, E.F. & McCoy, E.D. 1979. The statistics and biology of the species-area relationship. Amer. Nat. 113: 791-829.
Cornell, H.V. 1993. Unsaturated patterns in species assemblages: The role of regional processes in setting local species richness. pp. 243-252. In: Ricklefs, R.E. & Schluter, D. (eds), Species Diversity in Ecological Communities. Univ. Chicago Press, Chicago, London.
Courrier, K. 1992. Global Biodiversity Strategy - Guidelines for Action to Save, Study, and Use Earth's Biotic Wealth Sustainably and Equitably. IUCN, Gland.
Currie, D.J. 1991. Energy and large scale patterns of animal and plant species richness. Amer. Nat. 137: 27-49.
Dafni, A., Ivri, Y. & Brantjes, N.B.M. 1981. Pollination of *Serapias vomeracea* Briq. (Orchidaceae) by imitation of holes for sleeping solitary male bees (Hymenoptera). Acta Bot. Neerl. 30 (1/2): 69-73.
Darlington, P.J. 1957. Zoogeography. John Wiley, New York.
Dean, R.L. & Connell, J.H. 1987a. Marine invertebrates in algal succession. I. Tests of hypotheses to explain changes in diversity with succession. J. Exp. Mar. Biol. Ecol. 109: 217-247.
Dean, R.L. & Connell J.H. 1987b. Marine invertebrates in algal succession. II. Mechanisms linking habitat complexity with diversity. J. Exp. Mar. Biol. Ecol. 109: 249-273.
Diamond, J.M. 1972. Biogeographic kinetics: estimation of relaxation times for avifaunas of Southwest Pacific islands. Proc. Nat. Acad. Sci. USA 69: 3199-3203.
Diamond, J.M. 1973. Distributional ecology of New Guinea birds. Science 179: 59-769.
Diamond, J.M. 1988. Factors controlling species diversity: overview and synthesis. Ann. Missouri Bot. Gard. 75: 117-129.
Diamond, J.M. & Mayr, E. 1976. Species-area relation for birds of the Solomon Archipelago. Proc. Nat. Acad. Sci. USA 73 (1): 262-266.
Dierschke, H. 1994. Pflanzensoziologie: Grundlagen und Methoden. Ulmer Verlag, Stuttgart.
Douglas, M. & Lake, P.S. 1994. Species richness of stream stones: an investigation of the mechanism generating the species-area relationship. Oikos 69: 387-396.
During, H.J. *et al.* 1988. Diversity and pattern in plant communities. The Hague.
Ehrendorfer, F. 1962. Cytotaxonomische Beiträge zur Genese der mitteleuropäischen Flora und Vegetation. Ber. Dtsch. Bot. Ges. 75 (5): 137-152.
Eickwort, G.C. & Ginsberg H.S. 1980. Foraging and mating behavior in Apoidea. Ann. Rev. Entomol. 25: 241-246.
Elton, C. 1933. The Ecology of Animals. Methuen, London.
Eriksson, O. 1993. The species-pool hypothesis and plant community diversity. Oikos 68: 371-374.
Eriksson, O., Hansen, A. & Sunding, P. 1974. Flora of Macaronesia: Checklist of Vascular Plants. Univ. of Umeå, Sweden.

Evoy, W.H. & Jones, B. 1971. Motor patterns of male Euglossine bees evoked by floral fragrances. Anim. Behav. 19: 583-588.
Feeny, P. 1976. Plant apparency and chemical defence. Rec. Adv. Phytochem. 10: 1-40.
Forman, R.T.T., Galli, A.E. & Leck, C.F. 1976. Forest size and avian diversity in New Jersey woodlots with some land use implications. Oecologia 26: 1-8.
Frankel, O.H., Brown, A.H.D. & Burdon, J.J. 1995. The Conservation of Plant Biodiversity. Cambridge Univ. Press, Cambridge.
Gadgil, M. 1987. Diversity: cultural and biological. TREE 2 (12): 369-373.
Galli, A.E., Leck, C.F. & Forman, R.T. 1976. Avian distribution patterns in forest islands of different sizes in central New Jersey. Auk 93: 356-365.
Gaston, K.J. 1992. Regional numbers of insects and plant species. Funct. Ecol. 6: 243-247.
Gleason, H. 1926. The individualistic concept of the plant association. Bull. Torr. Bot. Club 53: 1-20.
Goetze, D. & Schwabe, A. 1997. Levels of biodiversity at different scales in space. Poster 40th Annual Symposium of the IAVS, Budweis 1997.
Goodall, D.W. 1952. Quantitative aspects of plant distribution. Biol. Rev. 27: 194-245.
Greig-Smith, P. 1964. Quantitative Plant Ecology. pp. 151-157. Butterworths, London.
Groombridge, B. 1992. Global Biodiversity - Status of Earth's Living Sources. A Report Compiled by the World Conservation Monitoring Centre. Chapman & Hall, London.
Günther, K. 1950. Ökologische und funktionelle Anmerkungen zur Frage des Nahrungserwerbs bei Tiefseefischen mit einem Exkurs über die ökologischen Zonen und Nischen. pp. 55-93. In: Grüneberg, H. & Ulrich, W. (eds), Moderne Biologie. Festschrift zum 60. Geburtstag von Hans Nachtsheim. Berlin.
Haber, W. 1978. Fragestellung und Grundbegriffe der Ökologie. pp. 74-79. In: Buchwald, K. & Engelhardt, W. (eds), Handbuch für Planung, Gestaltung und Schutz der Umwelt. Bd I. Die Umwelt des Menschen. München, Bern, Wien.
Haber, W. (in this volume). Conservation of biodiversity - scientific standards and practical realization.
Haeupler, H. 1997. Zur Phytodiversität Deutschlands: Ein Baustein zur globalen Biodiversitätsbilanz. Osnabrücker Naturw. Mitt. 23: 123-133.
Hamilton, T.H., Barth, R.H. Jr. & Rubinoff, I. 1964. The environmental control of insular variation in bird species abundance. Proc. Nat. Acad. Sci. USA 52: 132-140.
Hart, D.D. & Horwitz, R.J. 1991. Habitat diversity and the species-area relationship: alternative models and tests. pp. 47-68. In: Bell, S.S., McCoy, E.D. & Mushinsky, H.R. (eds), Habitat Structure: The Physical Arrangement of Objects in Space. Chapman & Hall, London.
Hawksworth, D.L. 1995. Biodiversity - Measurement and Estimation. Chapman & Hall, London.
Heywood, V.H. & Baste, I. 1995. Introduction. pp. 5-19. In: Heywood, V.H. & Watson, R.T. (eds), Global Biodiversity Assessment. Cambridge Univ. Press, Cambridge.
Heywood, V.H. & Watson, R.T. 1995. Global Biodiversity Assessment. Cambridge Univ. Press, Cambridge.
Hocking, B. & Sharplin, C.D. 1965. Flower basking by Arctic insects. Nature 4980: 215.
Huston, M.A. 1985. Patterns of species diversity on coral reefs. Ann. Rev. Ecol. Syst. 16: 149-177.
Huston, M.A. 1994. Biological Diversity - the Coexistence of Species on Changing Landscapes. Cambridge Univ. Press, Cambridge.
Johnson, M.P. & Raven, P.H. 1973. Species number and endemism: the Galapagos archipelago revisited. Science 179: 893-895.
Johnson, M.P. & Simberloff, D. 1974. Environmental determinants of island species numbers in the British Isles. J. Biogeogr. 1: 149-154.
Jurvik, J.O. & Austring, A.P. 1979. The Hawaiian avifauna: biogeographic theory in evolutionary time. J. Biogeogr. 6: 205-224.
Karr, J. 1981. Population variability and extinction in the avifauna of a tropical land bridge island. Ecology 63: 1975-1978.
Kevan, P.G. 1975. Sun-tracking solar furnaces in high Arctic flowers: significance for pollen and insects. Science 189: 723-726.
Kim, K.C. & Weaver, R.D. 1995. Biodiversity and Landscapes - a Paradox of Humanity. Cambridge Univ. Press, Cambridge.
Kohn, D.D. & Walsh, D.M. 1994. Plant species richness - the effect of island size and habitat diversity. J. Ecol. 82: 367-377.
Koepcke, H.W. 1971, 1973, 1974. Die Lebensformen. 2 Bde. Goecke und Evers, Krefeld.

Köppler, D. 1995. Vegetationskomplexe von Steppenheide-Physiotopen im Juragebirge. Diss. Bot. 249: 1-228. Cramer, Stuttgart, Berlin.
Köppler, D. & Schwabe, A. 1995. Typisierung und landschaftsökologische Gliederung S- und W-exponierter "Steppenheiden" mit Hilfe von Vegetationskomplexen. Ber. d. Reinh.-Tüxen-Ges. 8: 159-192.
Kratochwil, A. 1983. Zur Phänologie von Pflanzen und blütenbesuchenden Insekten (Hymenoptera, Lepidoptera, Diptera, Coleoptera) eines versaumten Halbtrockenrasens im Kaiserstuhl - ein Beitrag zur Erhaltung brachliegender Wiesen als Lizenzbiotope gefährdeter Tierarten. Beih. Veröff. Naturschutz Landschaftspflege Bad.-Württ. 34: 57-108.
Kratochwil, A. 1984. Pflanzengesellschaften und Blütenbesuchergemeinschaften: biozönologische Untersuchungen in einem nicht mehr bewirtschafteten Halbtrockenrasen (*Mesobrometum*) im Kaiserstuhl (Südwestdeutschland). Phytocoenologia 11 (4): 455-669.
Kratochwil, A. 1987. Zoologische Untersuchungen auf pflanzensoziologischem Raster - Methoden, Probleme und Beispiele biozönologischer Forschung. Tuexenia 7: 13-53.
Kratochwil, A. 1988a. Zur Bestäubungsstrategie von *Pulsatilla vulgaris* Mill. Flora 181: 261- 325.
Kratochwil, A. 1988b. Morphologische Untersuchungen im Blütenbereich in der Ontogenie von *Pulsatilla vulgaris* Mill. und ihre Bedeutung bei der Sippenabgrenzung. Bauhinia 9 (1): 15-26.
Kratochwil, A. 1989. Community structure of flower-visiting insects in different grassland types in Southwestern Germany (Hymenoptera Apoidea, Lepidoptera, Diptera). Spixiana 12 (3): 289-302.
Kratochwil, A. 1991a. Zur Stellung der Biozönologie in der Biologie, ihre Teildisziplinen und ihre methodischen Ansätze. Beih. 2 Verh. Ges. f. Ökologie: 9-44.
Kratochwil, A. 1991b. Blüten-/Blütenbesucher-Konnexe: Aspekte der Co-Evolution, der Co-Phänologie und der Biogeographie aus dem Blickwinkel unterschiedlicher Komplexitätsstufen. Annali di Botanica Vol IL: 43-108.
Kratochwil, A. 1996. Die Umweltkrise aus ökologischer Sicht - Historische Entwicklung und aktuelle Bilanz. pp. 7-152. In: Evangelische Akademie Baden (ed), Zukunft für die Erde, Bd 2: Dimensionen der ökologischen Krise. Herrenalber Protokolle 110.
Kratochwil, A. 1998. Zur Gültigkeit der Inseltheorie bei Festland-Ökosystemen - eine kritische Betrachtung auch für den Naturschutz. Braunschw. Geobot. Arb. 5: 7-37.
Kratochwil, A. & Klatt, M. 1989a. Apoide Hymenopteren der Stadt Freiburg i. Br. - submediterrane Faunenelemente an Standorten kleinräumig hoher Persistenz. Zool. Jb. Syst. 116: 379-389.
Kratochwil, A. & Klatt, M. 1989b. Wildbienengemeinschaften (Hymenoptera Apoidea) an spontaner Vegetation im Siedlungsbereich der Stadt Freiburg. Braun-Blanquetia 3 (2): 421-438.
Kratochwil, A. & Aßmann, T. 1996a. Biozönotische Konnexe im Vegetationsmosaik nordwestdeutscher Hudelandschaften. Ber. d. Reinh.-Tüxen-Ges. 8: 237-282.
Kratochwil, A. & Aßmann, T. 1996b. Biozönologische Untersuchungen in Hudelandschaften des nordwestdeutschen Tieflandes. Verh. Ges. f. Ökologie 26: 229-237.
Krattinger, A.F., McNeely J.A., Lesser, W.H., Miller, K.R., Hill, Y.St. & Senanayke, R. (eds). 1994. Widening Perspectives on Biodiversity. IUCN, Gland.
Krebs, C.J. 1989. Ecological Methodology. Harper & Row, New York, Cambridge.
Kugler, H. 1970. Einführung in die Blütenökologie. 2nd ed. Fischer Verlag, Stuttgart.
Kullenberg, B. 1961. Studies on *Ophrys* pollination. Zool. Bidr. Uppsala 34: 1-340.
Lack, D.L. 1947. Darwin's Finches. Cambridge Univ. Press, Cambridge.
Lack, D.L. 1969. The number of bird species on islands. Bird Study 16: 193-209.
Lassen, H.H. 1975. The diversity of freshwater snails in view of the equilibrium theory of island biogeography. Oecologia 19: 1-8.
Latham, R.E. & Ricklefs, R.E. 1993. Global patterns of tree species richness in moist forests: energy-diversity theory does not account for variation in species richness. Oikos 67: 325-333.
Lawton, J.H. & Brown, V.K. 1993. Redundancy in ecosystems. pp. 255-270. In: Schulze, E.-D. & Mooney, H.A. (eds), Biodiversity and Ecosystem Function. Ecological Studies 99. Springer Verlag, Berlin, Heidelberg, New York.
Lawton, J.H. 1990. Species richness and population dynamics of animal assemblages. Pattern in body size: abundance space. Phil. Trans. Roy. Soc. London B 330: 283-291.
Lindroth, C.H. 1960. The ground-beetles of the Azores (Coleoptera: Carabidae) with some reflexions on over-seas dispersal. Bol. Mus. Muni. Funchal XIII, Art. 31: 5-48.
Lugo, A. E. 1992. Schätzungen des Rückgangs der Artenvielfalt tropischer Wälder. pp. 76-89. In: Wilson, E.O. (ed), Ende der biologischen Vielfalt? Der Verlust von Arten, Genen und Lebensräumen und die Chancen für eine Umkehr. Springer Verlag, Heidelberg, Berlin, New York.

Lynch, J.F. & Wigham, D.F. 1984. Effects of forest fragmentation on breeding bird communities in Maryland, USA. Biol. Conserv. 28: 287-324.
MacArthur, R.H. & Wilson, E.O. 1967. The Theory of Island Biogeography. Princeton Univ. Press, Princeton, NJ.
Magurran, A.E. 1988. Ecological Diversity and Its Measurement. Princeton Univ. Press, Princeton, NJ.
Matthies, D., Schmid, B. & Schmid-Hempel, P. 1995. The importance of population process for the maintenance of biological diversity. GAIA 4 (4): 199-209.
May, R.M. 1975. Pattern of species abundance and diversity. pp. 81-120. In: Cody, M.L. & Diamond, J.M. (eds), Ecology and Evolution of Communities. Harvard Univ. Press, Cambridge, Mass.
May, R.M. 1978. Dynamics and diversity of insect faunas. pp. 188-204. In: Mound, L.A. & Waloff, N. (eds), Diversity of Insect Fauna. Blackwell Scientific, Oxford.
May, R.M. 1988. How many species are there on earth? Science 241: 1441-1449.
Mayer, D.F. & Johansen C.A. 1978. Bionomics of *Meloe niger* Kirby (Coleoptera: Meloidae), a predator of the Alkali bee, *Nomia melanderi* Cockerell (Hymenoptera: Halictidae). Melanderia: 1-22.
Mayr, E. 1943. The zoogeographic position of the Hawaiian Island. Condor 45: 45-48.
McNeely, J., Miller, K.A., Reid, W.V., Mittermeier, R.A. & Werner, T.B. 1990. Conserving the World's Biological Diversity. IUCN, Gland, Washington.
Morse, D.H. 1984. How crab spiders (Araneae, Thomisidae) hunt at flowers. J. Arachnol. 12: 307-316.
Myers, N. 1985. Die sinkende Arche: bedrohte Natur, gefährdete Arten. Westermann Verlag, Braunschweig.
O'Connor, N.A. 1991. The effects of habitat complexity on the macroinvertebrates colonizing wood substrates in a lowland stream. Oecologia 85: 504-512.
Oberdorfer, E. 1994. Pflanzensoziologische Exkursionsflora. 7th ed. Ulmer Verlag, Stuttgart.
Odum, E.P. 1971. Fundamentals of Ecology. Saunders Comp., London, Philadelphia, Toronto.
Odum, E.P. 1983. Grundlagen der Ökologie. Thieme Verlag, Stuttgart, New York.
Osche, G. 1966. Grundzüge einer allgemeinen Phylogenetik. pp. 817-906. Handbuch der Biologie III/2. Akad. Verlagsges., Frankfurt.
Osche, G. 1975. Die vergleichende Biologie und die Beherrschung der Mannigfaltigkeit. BIUZ 5: 139-146.
Palmgreen, A. 1925. Die Artenzahl als pflanzengeographischer Charakter. Fennia 46 (2): 1-144.
Paulus, H.F. 1978. Co-Evolution zwischen Blüten und ihren tierischen Bestäubern. Sonderband naturwiss. Ver. Hamburg 2: 51-81.
Paulus, H.F. 1988. Co-Evolution und einseitige Anpassungen in Blüten-Bestäubersystemen: Bestäuber als Schrittmacher in der Blütenevolution. Verh. Dtsch. Zool. Ges. 81: 25-46.
Pickett, S.T.A. & White P.S. 1985. The Ecology of Natural Disturbance and Patch Dynamics. Academic Press, New York.
Pignatti, S. 1983. Human impact on the vegetation of the Mediterranean Basin. pp. 151-161. In: Holzner, W., Werger, M.J.A. & Ikusima, I. (eds), Man's Impact on Vegetation. Dr W. Junk Publ., The Hague.
Pignatti, G. & Pignatti, S. (in this volume). Biodiversity in Mediterranean ecosystems.
Pott, R. 1995. Die Pflanzengesellschaften Deutschlands. 2nd ed. Ulmer Verlag, Stuttgart.
Pott, R. (in this volume). Diversity of pasture-woodlands of north-western Germany.
Pott, R. & Hüppe, J. 1991. Die Hudelandschaften Nordwestdeutschlands. Abhandl. Westfäl. Mus. Naturkde Münster 53 (1/2): 1-313.
Preston, F.W. 1962. The canonical distribution of commonness and rarity. Ecology 43: 185-215, 410-432.
Rabeler, W. 1957. Die Tiergesellschaft eines Eichen-Birkenwaldes im nordwestdeutschen Altmoränengebiet. Mitt. flor. soz. Arbeitsgem. N.F. 6/7: 297-319.
Rabeler, W. 1960. Die Artenbestände der Regenwürmer in Laubwald-Biozönosen (Querco-Fagetea) des oberen und mittleren Wesergebietes. Mitt. Flor.-soz. Arbeitsgem. N.F. 8: 333-337.
Raunkiaer, C. 1907/1937. Planterigets Livsformer og deres Betydning for Geografien. Kjöbenhavn, Kristiana. Translation into English: Gilbert-Carter, H. 1937: Plant Life-forms. Clarendon Press, Oxford.
Raustiala, K. & Victor, D.G. 1996. Biodiversity since Rio: The Future of the Convention on Biological Diversity. Environment (Washington) 38: 17-20, 37-43.
Reaka-Kudla, M.L., Wilson, D.E. & Wilson, E.O. (eds). 1997. Biodiversity II. Understanding and Protecting Our Biological Resources. Joseph Henry Press, Washington D.C.
Reed, T. 1981. The number of breeding landbird species on British islands. J. Anim. Ecol. 50: 613-624.
Remmert, H. (ed). 1991. The Mosaic-Cycle Concept of Ecosystems. Ecological Studies 85. Springer, Berlin, Heidelberg, New York.
Ricklefs, R.E. & Cox, G.W. 1972. The taxon cycle in the land bird fauna of the West Indies. Amer. Nat. 106: 195-219.

Ricklefs, R.E. & Schluter, D. (eds). 1993. Species Diversity in Ecological Communities. Historical and Geographical Perspectives. Univ. of Chicago Press, Chicago, London.
Ricklefs, R.E. 1987. Community diversity: Relative roles of local and regional processes. Science 235: 167-171.
Rosenzweig, M.L. 1995. Species Diversity in Space and Time. Cambridge Univ. Press, Cambridge.
Schaefer, M. (in this volume). The diversity of the fauna of two beech forests: some thoughts about possible mechanisms causing the observed patterns.
Schoener, T.W. (1986). Resource partitioning. pp. 91-126. In: Kikkawa, J. & Anderson, D.J. (eds), Community Ecology. Pattern and Process. Blackwell, Melbourne.
Schulze, E.-D. & Mooney, H.A. (eds). 1993. Biodiversity and Ecosystem Function. Ecological Studies 99. Springer, Berlin, Heidelberg, New York.
Schwabe, A. 1990. Stand und Perspektiven der Vegetationskomplex-Forschung. Ber. d. Reinh.-Tüxen-Ges. 2: 45-60.
Schwabe, A. 1991a. A method for the analysis of temporal changes in vegetation pattern at the landscape level. Vegetatio 95: 1-19.
Schwabe, A. 1991b. Perspectives of vegetation complex research and bibliographic review of vegetation complexes in vegetation science and landscape ecology. Excerpta Botanica 28 (Sect. B.): 223-243.
Schwabe, A. 1995. *Kochia prostrata* (L.) Schrader-reiche Pflanzengesellschaften und Vegetationskomplexe unter besonderer Berücksichtigung des Aostatales. Carolinea 53: 83-98.
Schwabe, A. (in this volume). Spatial arrangements of habitats and biodiversity: an approach to a sigmasociological view.
Schwabe, A. & Mann, P. 1990. Eine Methode zur Beschreibung und Typisierung von Vogelhabitaten, gezeigt am Beispiel der Zippammer (*Emberiza cia*). Ökologie der Vögel (Ecology of Birds) 12: 127-157.
Schwabe, A., Köppler, D. & Kratochwil, A. 1992. Vegetationskomplexe als Elemente einer landschaftsökologisch-biozönologischen Gliederung, gezeigt am Beispiel von Fels- und Moränen-Ökosystemen. Ber. d. Reinh.-Tüxen-Ges. 4: 135-145.
Schwabe, A. & Kratochwil, A. 1994. Gelten die biozönotischen Grundprinzipien auch für die landschaftsökologische Dimension? Einige Überlegungen mit Beispielen aus den Inneralpen. Phytocoenologia 24: 1-22.
Simberloff, D.S. 1974. Equilibrium theory of island biogeography and ecology. Ann. Rev. Ecol. Syst. 5: 161-182.
Simberloff, D. 1976. Experimental zoogeography of islands: effects of island size. Ecology 57: 629-648.
Simberloff, D. & Cox, J. 1987. Consequences and costs of conservation corridors. Conserv. Biol. 1: 63-71.
Simberloff, D., Farr, J.A., Cox, J. & Mehlmann, D.W. 1992. Movement corridors: Conservation bargains or poor investments? Conserv. Biol. 6: 493-505.
Smith, F.D.M., May, R.M., Pellew, R., Johnson, T.H. & Walter, K.R. 1993. How much do we know about the current extinction rate? TREE 8: 375-378.
Solbrig, O.T. 1991. Biodiversity. Scientific Issues and Collaborative Research Proposals. UNESCO, Paris.
Solbrig, O.T. 1994. Biodiversität. Wissenschaftliche Fragen und Vorschläge für die internationale Forschung. Bonn: Deutsche UNESCO-Kommission (MAB).
Solbrig, O.T., van Emden, H.M. & van Oordt, P.G.W.J. 1992. Biodiversity and Global Change. Paris, International Union of Biological Science (IUBS), Monograph No. 8.
Stearns, S.C. *et al.* 1990. Biodiversity. pp. 46-74. In: Schweizerischer Wissenschaftsrat (ed), Forschungspolitische Früherkennung. Technologien zur Erhaltung der biologischen Vielfalt.
Strasburger, E. 1991. Lehrbuch der Botanik. 33rd ed. (edited by P. Sitte, H. Ziegler, F. Ehrendorfer, A. Bresinsky). Gustav Fischer Verlag, Stuttgart.
Tangney, R.S., Wilson, J.B. & Mark, A.F. 1990. Bryophyte island biogeography: a study of Lake Manapouri, New Zealand. Oikos 59: 21-26.
Terborgh, J. 1974. Faunal equilibria and the design of wildlife preserves. In: Golley, F. & Medina, E. (eds), Tropical Ecological Systems: Trends in Terrestrial and Aquatic Research. Springer Verlag, New York.
Thienemann, A. 1920. Die Grundlagen der Biocoenotik und Monards faunistische Prinzipien. Festschrift für Zschokke No. 4: 1-14.
Thienemann, A. 1956. Leben und Umwelt. Vom Gesamthaushalt der Natur. Hamburg.
Tischler, W. 1949. Grundzüge der terrestrischen Tierökologie. Vieweg & Sohn, Braunschweig.
Trepl, L. 1988. Gibt es Ökosysteme? Landschaft & Stadt 20 (4): 176-185.
Trepl, L. 1994. Geschichte der Ökologie vom 17. Jahrhundert bis zur Gegenwart. 2nd ed. Beltz Athenäum, Studienbücher der Geschichte.

UNEP (United Nations Environmental Programme). 1993. Global Biodiversity. UNEP/GEMS Environment Library No. 11- Nairobi: UNEP.
UNEP (United Nations Environmental Programme). 1995. Global Biodiversity Assessment. Cambridge Univ. Press, Cambridge.
Van der Maarel, E. 1988. Species diversity in plant communities in relation to structure and dynamics. pp. 1-14. In: During, H.J. *et al.* (eds), Diversity and Pattern in Plant Communities.
Van Leeuwen, C.G. 1966. A relation-theoretical approach to pattern and process in vegetation. Wentia 15: 25-46.
Vestal, A.G. 1949. Minimum areas for different vegetations: their determination from species-area curves. Univ. Ill. Biol. Monogr. 20 (1).
Vestal, A.G. & Heermans, M.F. 1945. Size requirements for reference areas in mixed forests. Ecology 26: 122-134.
Vida, G. 1994. Global issues of genetic diversity. pp. 9-19. In: Loeschke, V. Tomiuk, J. & Jain, S.K. (eds), Conservation Genetics. Birkhäuser, Basel.
Vogel, S. 1988. Die Ölblumensymbiosen - Parallelismus und andere Aspekte ihrer Entwicklung in Raum und Zeit. Z. zool. Syst. Evol.-Forsch. 26: 341-362.
Walker, B.H. 1992. Biodiversity and ecological redundancy. Conserv. Biol. 6: 8-23.
Walter, H. & Breckle, S.-W. 1984. Ökologie der Erde. Bd 2: Spezielle Ökologie der Tropischen und subtropischen Zonen. Fischer Verlag, Stuttgart.
Walter, H. & Straka, H. 1970. Arealkunde. Floristisch-historische Geobotanik. Ulmer Verlag, Stuttgart.
Westrich, P. 1989. Die Wildbienen Baden-Württembergs. Allgemeiner Teil: Lebensräume, Verhalten, Ökologie, Schutz. Ulmer Verlag, Stuttgart.
Williams, C.B. 1953. The relative abundance of different species in a wild animal population. J. Anim. Ecol. 22: 14-31.
Wilson, E.O. 1961. The nature of the taxon cycle in the Melanesian ant fauna. Amer. Nat. 95: 169-193.
Wilson, E.O. 1988. Biodiversity. National Acad. Press, Washington.
Wilson, E.O. 1989. Die Bedrohung des Artenreichtums. Spektr. Wissensch. 11: 88-95.
Wilson, E.O. 1992. The Diversity of Life. Harvard Univ. Press, Cambridge.
Wissenschaftlicher Beirat der Bundesregierung "Globale Umweltveränderungen". 1993. Grundstruktur globaler Mensch-Umwelt-Beziehungen; Jahresgutachten 1993. Economica Verlag, Bonn.
Whittaker, R.H. 1972. Evolution and measurement of species diversity. Taxon 21 (2/3): 213-251.
Whittaker, R.H. 1975. Communities and Ecosystems. 2nd ed. New York, London.
Whittaker, R.H. 1977. Evolution of species diversity in land communities. Evol. Biol. 10: 1-67.
World Resources Institute (WRI) 1992. Global Biodiversity Strategy. Washington D.C. USA: WRI.

THE DIVERSITY OF THE FAUNA OF TWO BEECH FORESTS: SOME THOUGHTS ABOUT POSSIBLE MECHANISMS CAUSING THE OBSERVED PATTERNS

MATTHIAS SCHAEFER

II. Zoologisches Institut, Abteilung Ökologie, Universität Göttingen, Berliner Straße 28, D-37075 Göttingen, Germany

Key words: food web, insects, invertebrates, regional species pool, species number, trophic organisation

Abstract

The diversity of the animal community was studied for two North German beech forests with different humus types (mull and mor). The number of species was high: 1918 in the mull-structured beech forest (915 without Diptera and Hymenoptera), 694 in the mor-structured beech forest (with the Diptera and Hymenoptera not analysed). Ecological mechanisms causing diversity relate to life-form characteristics of the species in a taxon, the regional species pool, and habitat features.

Small body size and short life-cycles of the species are a prerequisite for high diversity in a taxon. Among the five trophic guilds differentiated for the analysis, predators and parasitoids clearly dominated, the ranking for the other guilds was phytophages > saprophages > microphytophages. For comparable groups species numbers in the forest (the local species pool) were positively correlated with regional species diversity (taking Central Europe as the regional species pool). Phytophages occur in the forest, however, their proportion in the species pool is low, whereas a high percentage of Central European parasitoids is represented in the local pool.

A set of hypotheses is discussed that might explain high diversity of a taxon in the local community. For the two forests in question the higher species diversity in the mull-structured site could be explained by the more heterogeneous environment, the less extreme abiotic factors, disturbance caused by the bioturbating macrofauna in the litter-soil system, and more intense biotic interactions.

Introduction

Life is diverse. Discussion about "biodiversity" has become fashionable recently, because man causes extinction of species to a large extent (Wilson 1992; Smith *et al.* 1993). Another reason for the recent prevalence of discussion on diversity (*e.g.* Ricklefs &

A. Kratochwil (ed.), Biodiversity in ecosystems, 39-57

Schluter 1993; Huston 1994; Rosenzweig 1995) is the interest in the organisation of communities and the structure of food webs, both normally based on a high number of species. Additionally, a tantalizing question is that on the function of diversity for ecosystem processes (Lawton & Brown 1993; Lawton 1994; Jones *et al.* 1994), among others providing arguments for the preservation of species (Wilson 1992; Walker 1992).

Diversity denotes the manifold expressions of life from genetic diversity to species numbers and biology of the species to the organisation of communities, ecosystems, landscapes and the whole biosphere. In a strict sense it describes two components of a community: species number and the relative abundance of the populations. Both aspects are combined in diversity indices (Krebs 1989). In ecosystem studies three levels of diversity can be analysed: diversity of the elements and interactions ("pattern"), "mechanisms" causing this pattern, and the role of diversity for the system ("processes"), such as predation pressure, energy flow or chemical flow.

In my contribution about diversity of beech forests I will restrict myself to species numbers and the specific biology of the species, regardless whether the species is abundant or rare. Based on long-term studies of a mull-structured and a mor-structured beech forest in Northern Germany I will discuss the following questions and hypotheses.

1. Are life-form attributes of the species related to the diversity of the community? I will pay special attention to the nutritional type of the animal groups.
2. Does the regional species pool (the fauna of Central Europe) determine local diversity in the forest?
3. Which ecological factors, such as heterogeneity of the environment or disturbance of the habitat, contribute to high faunal diversity?

The approach raises conceptual and methodological problems: As only two forests are compared, generalisations about the importance of habitat characteristics are not possible. Instead, I focus on the analysis of different animal taxa within one forest; the "study unit" must comprise species belonging to the same guild or a similar guild. Often factors are linked with one another and cannot be disentangled, *e.g.* detritus feeders (variable "trophic type") are simultaneously litter and/or soil dwellers (variable "spatial distribution or stratification"). Because the variables cannot be quantified, a statistical analysis is not possible. Hence, relations between diversity attributes and possible factors responsible for the observed patterns often are mere guesses. My aim is to present some thought-provoking ideas as a basis for future studies.

Case study: the beech forests "Göttinger Wald" and "Solling"

The beech forest Göttinger Wald is situated in southern Lower Saxony (West Germany) on a plateau of Muschelkalk with an elevation of about 420 m above sea level. The beech forest about 8 km east of Göttingen is approximately 115-125 years old and has a rather uniform canopy layer consisting almost exclusively of beech (*Fagus sylvatica* L.) trees, which form a dense crown layer. Some trees of other species, mostly ash (*Fraxinus excelsior* L.) and maple (*Acer platanoides* L.), and more rarely *Acer pseudoplatanus* L., *Quercus robur* L., *Q. petraea* (Matt.) Liebl. and *Ulmus scabra* Mill., are interspersed. A shrub layer has not developed, though in gaps a enhanced growth of young

F. excelsior trees can be found. The herb layer is, with few exceptions, dense and diverse. Dominant spring geophytes are *Allium ursinum* L. and *Anemone nemorosa* L.; additional dominant herbs are *Asarum europaeum* L., *Hordelymus europaeus* (L.) Jess., *Galium odoratum* (L.) Scop., *Lamiastrum galeobdolon* (L.) Nath., *Mercurialis perennis* L., *Oxalis acetosella* L. and *Primula elatior* (L.) Hill. The forest belongs to the Melico-Fagetum hordelymetosum. The soil, which is shallow and belongs to the soil series on carbonate rocks, consists of "terra fusca-rendzina" (about 50%), rendzina (about 26%), terra fusca (about 14%), and some other modifications containing brown earth. The organic layer comprised leaf material of a mean annual mass of 550 g dry mass m^{-2}; annual canopy leaf litter fall amounted to 309 g dry mass m^{-2} (mean value for 1981-1991, Schmidt, unpublished results). The pH ranges from 6.8 to 4.3 with a mean value of about 5.8. The cation exchange capacity is high. The average annual temperature is about 7 °C, the mean annual precipitation amounts to approximately 700 mm. Further details are given by Schaefer (1990, 1991) and Schaefer & Schauermann (1990).

The Solling beech forest is located on the plateau of the High Solling, a mountain range of medium altitude (500 m above sea level) about 55 km northwest of Göttingen. It is a pure beech (*F. sylvatica*) stand without a shrub and herb layer, and can be classified as a Luzulo-Fagetum. Parent material of this acidophilous forest is a loess creeping earth of 40-60 cm thickness above a paraautochthonic creeping earth consisting of loamy weathered Buntsandstein material. The unweathered Buntsandstein of Triassic age begins at depths of about 100 cm. Hence the soil belongs to the series on carbonate-free silicate rocks: it is an acid brown earth with a moder humus form. The mass of organic layer on the soil increased from 2960 g m^{-2} in the year 1966 to 4460 g dry mass m^{-2} in 1979; annual canopy leaf litter fall had a mean value of 290 g dry mass m^{-2} for the period from 1967 to 1976. pH values range from about 3 to 4. The cation exchange capacity is low. The mean annual precipitation amounts to 1063 mm. Further details are given in Ellenberg *et al.* (1986).

The fauna of the two forests was being sampled with varying intensities for a long time, from 1980 to 1995 for the Göttinger Wald, from 1995 to 1997 for the Solling. For many (not all) animal groups, species numbers of the two forests can be compared because the intensity of sampling was similar. A synopsis of species numbers, population density and biomass is presented in Table 1, which is based on the data collection published in Schaefer & Schauermann (1990). Recent results are given for Nematoda (Solling) (Alphei 1995), Cryptostigmata, Gamasina and Uropodina (Göttinger Wald) (Schulz 1991), Gamasina (Solling) (Buryn, unpublished results) and Diptera (Göttinger Wald) (Hövemeyer 1992). Some minor groups have been omitted. For the Solling forest several important and species-rich taxa have not been studied. The mull-structured forest is characterized by high faunal biomass with earthworms as the dominant macrofauna group. This saprophagous guild is responsible for a high degree of bioturbation. In the mor-structured forest microphytophagous mesofaunal taxa (oribatid mites, collembolans) are prevalent. Generally, species diversity is lower in the Solling forest. A notable exception are staphylinid beetles. The total animal species number in the forest "Göttinger Wald" amounts to 1918, among them 704 hymenopteran and 299 dipteran species; the forest "Solling" contains 734 species (with the Hymenoptera and Diptera not studied), well below the 915 species of the mull-structured forest (cf. Table 4).

Table 1. Synopsis of species richness (S), mean annual population density (N, ind m^{-2}) and mean annual biomass (B, mg dry mass m^{-2}) in the beech forests Göttinger Wald (with mull soil) and Solling (with moder soil). Modified from Schaefer & Schauermann (1990). Some further groups are mentioned in Table 4

Animal Group	Göttinger Wald			Solling		
	S	N	B	S	N	B
Microfauna						
Flagellata	?	2.7·109	54	?	?	?
Amoebina	?	3.5·109	1,133	?	?	?
Testacea	65	84·106	343	51	57·106	256
Turbellaria	3	859	8	3	1,882	4
Nematoda	65	732,000	146	90	3·106	65
Rotatoria	13	4,893	5	?	?	?
Tardigrada	4	4,207	4	?	41	9
Harpacticoida	?	3,873	2	1	3,300	0.6
Saprophagous and microphytophagous mesofauna						
Enchytraeidae	36	22,300	600	15	108,000	1,640
Cryptostigmata	75	22,445	241	72	101,810	195
Uropodina	11	1,971	19	4	1525	?
Symphyla	2	57	?	1	?	?
Diplura	?	161	?	>1	277	?
Protura	?	2,481	?	>1	278	?
Collembola	48	37,835	153	50	63,000	246
Zoophagous mesofauna						
Gamasina	80	3,151	50	53	10,8	397
Saprophagous macrofauna						
Gastropoda	30	120	430	4	0	0
Lumbricidae	11	205	10,700	4	19	168
Isopoda	6	286	93	0	0	0
Diplopoda	6	55	618	1	0	0
Elateridae	11	$37^{*)}$	$104^{*)}$	4	$332^{*)}$	$706^{*)}$
Diptera	299	$2,843^{*)}$	$161^{*)}$	?	$7,415^{*)}$	$628^{*)}$
Zoophagous macrofauna						
Araneida	102	140	47	93	462	173
Pseudoscorpionida	3	35	16	2	89	10
Opilionida	8	19	11	4	20	6
Chilopoda	10	187	265	7	74	155
Carabidae	24	5	144	26	7	93
Staphylinidae	85	103	76	117	314	180

? = not studied; 0 = not present; *) = larvae.

Diversity in the forest does not only concern species richness. Spatial distribution, life-cycles and nutritional biology of the different animal groups vary to a high degree. This is especially true for the utilization of food resources. For instance, Hövemeyer (1995) differentiates between 16 trophic types for the dipterans of the forest "Göttinger Wald".

The regional species pool

I regard Central Europe as the regional species pool, a source for populations colonizing the forests. The region is inhabited by about 40,000 species, most of them are Arthropoda. Prominent groups are beetles and hymenopterans (Table 2).

Table 2. Species numbers (S) of animal taxa in Central Europe (CE). Only those groups are considered here that are discussed in this context. Literature as a general source of information about species numbers: Dunger (1983), Jacobs & Renner (1988), Hannemann *et al.* (1989, 1994), Schaefer (1994)

Animal group	S_{CE}	Additional literature
Protozoa		
Testacea	900	
Turbellaria	150	
Nematoda	1100	
Rotatoria	600	
Annelida		
Enchytraeidae	100	Wilcke (1963)
Lumbricidae	70	Wilcke (1963)
Gastropoda	400	Kerney *et al.* (1983)
Arthropoda		
Tardigrada	60	
Araneida	900	Heimer & Nentwig (1991)
Pseudoscorpionida	30	
Opilionida	40	Martens (1978)
Gamasina	1000	Karg (1993)
Uropodina	184	Karg (1989)
Cryptostigmata	447	Weigmann (1981)
Isopoda	40	
Chilopoda	50	
Diplopoda	130	
Collembola	300	
Heteroptera	800	Günther & Schuster (1990)
Auchenorrhyncha	500	
Aphidina	850	
Coleoptera	8000	Kloet *et al.* (1977); Lucht (1987)
Carabidae	500	Lucht (1987)
Staphylinidae	1300	Lucht (1987)
Elateridae	150	Lucht (1987)
Chrysomelidae	500	Lucht (1987)
Curculionidae	1000	Lucht (1987)
Hymenoptera	10000	Fitton *et al.* (1978)
Diptera	8000	Kloet & Hincks (1976); Hövemeyer (1995)
Lepidoptera	3000	

When calculating the species numbers, several problems arise. The geographic definition of Central Europe is not precise; the Alps with alpine species should not be included, but where set the border? Groups with a large proportion of limnic species (*e.g.* Testacea) represent a smaller regional pool for the forest as a terrestrial habitat than indicated by the number. For some taxa (*e.g.* Nematoda, Diptera, Hymenoptera) the number of Central European species is not known exactly. I have to present rough estimates. I have utilized as important sources of information Dunger (1983), Jacobs & Renner (1988), Hannemann *et al.* (1992, 1994) and Schaefer (1994): other publications relevant for the group in question are presented in Table 2.

Patterns of diversity

The analysis of the patterns of diversity can be based on taxa or on the different life-forms.

Taxonomic diversity

Among others, the following animal groups in Central Europe are highly diverse: nematodes, spiders, gamasine mites, hemipterans, beetles and dipterans. Obviously some groups are privileged, as they can attain high diversity values. Trivial interpretations for high speciation rates can be found in the literature; some examples: mites have a small body size and meet a coarse-grained environment; insects are able to fly and have a high dispersal power; beetles have a hard exoskeleton serving as defence against predators; spiders produce silk for catching prey and for dispersal. The pattern of diversity in Central Europe roughly corresponds to the world-wide pattern of species numbers (Wilson 1992). A notable exception are the hymenopteran parasitoids which reach a higher proportion of the fauna in the temperate region of Europe. The representation of the taxa in the forest at least partly mirrors their relative occurrence in Central Europe (see Table 4).

Diversity of life forms

As indicated in the foregoing paragraph, some diversity patterns emerge for the life-form types of the fauna in the forests. These patterns are plausible from an intuitive perspective. Species with small body size and short life-cycles dominate in numbers (May 1978; Rosenzweig 1995). Species with a long life-cycle and high longevity (*e.g.* Diplopoda, Chilopoda) constitute a minority. I refrain from documenting the species size and life-cycle distribution for the fauna in the forests in detail.

Diversity in the vegetation layer and in the mineral soil with their extreme environments is lower than in the litter layer and the ground zone. For many groups (*e.g.* Araneida, Collembola, Coleoptera) a maximum of species numbers can be observed in the litter and lower herb stratum. However, heterogeneities in vertical distribution depend on the taxon in question. Earthworms are soil-dwellers, and lepidopterans have nothing to do with the ground zone.

I differentiate between five trophic categories: phytophages, saprophages, microphytophages and zoophages (with predators and parasitoids). Regarding only the diversity of food resources, some possible patterns emerge. Zoophages are more diverse than phytophages, saprophages or microphytophages. Phytophages are less diverse than saprophages plus microphytophages. Saprophages may equal microphytophages in species numbers. Primary decomposers are less species-rich than secondary decomposers, mycophages probably are present in numbers comparable to bacteriophages. The diversity of parasitoids (with many specialists) surpasses that of predators (with many generalists) (see hypotheses 6 and 15 below).

Mechanisms causing diversity

Two principal types of mechanisms exist that determine the diversity of a taxon in a given habitat (Eriksson 1993). During evolution high speciation rates and/or low extinction rates enhance diversity; in situations of low rates of species formation and frequent extinction diversity reaches only low levels. In ecological time the animal community is composed of the existing set of species which colonize the habitat, build up populations and are capable of persisting for a certain time.

Effects in evolutionary time

Two models may apply to effects on diversity during evolution (Ricklefs 1990).

The higher the age of the community the more species have originated; according to this "time hypothesis" (Latham & Ricklefs 1993) older habitats should contain more species. This hypothesis can be tested by comparing forests of different ages. Indeed, Terrell-Nield (1990) found that the carabid fauna of old forests is richer in species than that in comparable younger forests. However, two reservations have to be made. The continuous rise of species numbers is only possible if the community has "empty niches" and tolerates species additions (see below) (Cornell 1993; Cornell & Lawton 1992). Larger numbers of species in older habitats may be explained by a colonization process from a species pool still going on. This latter explanation would only apply in cases where the filling up of a habitat with species from outside sources requires a long time. However, there are indications that colonization processes are rapid; Rosenzweig (1995) regards a period of about 1,000 years as typical for the completion of colonization processes. The Solling forest is much older than the forest "Göttinger Wald"; however, no effects of time on community patterns are apparent.

The number of species present in a given community is the result of speciation and extinction. Under equilibrium conditions both rates are equal ("equilibrium model"). No methods can be imagined to test this model, because it is not known if the species in a community evolved together. Some authors take this for granted and develop evolutionary explanations for the composition of the community, assuming an overall importance of interspecific competition which leads to an extinction of populations and character divergence between coexisting populations (Cracraft 1994). Such an approach applied to forest habitats, *e.g.* by Müller (1985) for Carabidae, can be pure speculation.

Effects in ecological time

The composition of the animal community is controlled by three sets of ecological factors (Fig. 1, Table 3).

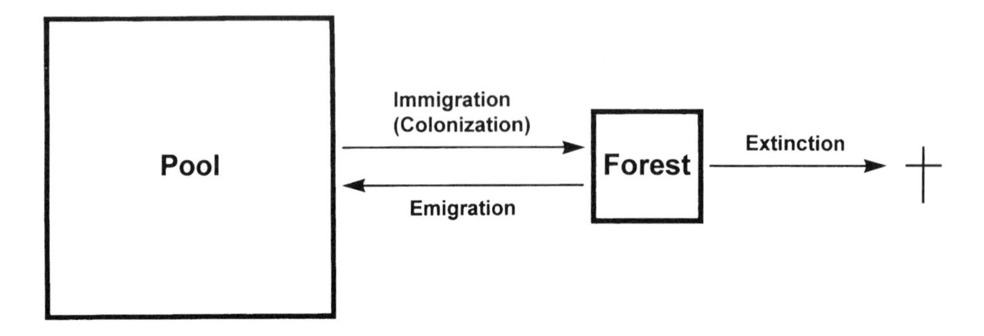

Figure 1. The exchange of species between the species pool and a forest habitat.

Table 3. Factors influencing the size of the local species pool

Regional pool	—— Colonization ——►	Local pool
Life form		
	Mobility	Body size Life-cycle Degree of euryoecy Degree of silvicoly Trophic type
Pool size		
Species number		
Extrinsic factors		
Distribution of sources	Factors favouring dispersal	Special habitat features Habitat favourableness Extremity of habitat factors Spatial heterogeneity Temporal heterogeneity Diversity of food resources Predation Interspecific competition Productivity Disturbance Stochastic variation

One determinant is the size of the regional species pool. This influential variable is modified by the distribution pattern of habitat patches serving as sources for the community of the forest.

Important are the dispersal characteristics of the species belonging to it, summarily denoted as "mobility". External factors, such as frequent wind, may enhance colonization rates.

Factors modulating the ability of members of a pool population to settle in the forest are life-form characteristics, among others: body size, length of life-cycle, preferred stratum, degree of euryoecy, degree of silvicoly (intrinsic factors). Extrinsic factors concern features of the habitat enabling the colonizing species to settle in the forest. Important habitat characteristics are abiotic habitat factors in a continuum from favourableness to unfavourableness, spatial and temporal heterogeneity, heterogenity of food resources, diversity of biotic factors such as predation or interspecific competition, productivity, stochastic variation, disturbance. The relations between populations and surroundings can be regarded as niche dimensions.

One reservation has to be made: the factors "dispersal and migration" and "ability of successful colonization" are only important if the community is not saturated, that is if there are species that would be capable of living in the forest but have not yet arrived ("time hypothesis").

In the following paragraphs I will discuss some of these conditions possibly influencing the composition and diversity of the fauna in the two forests in more detail. For the sake of clarity and precision, I will do this by formulating hypotheses.

Ecological hypotheses about diversity in the forests

Hypothesis 1: Diversity is higher for taxa with small species

This clearly is the case; the minority of species in both forests is of larger body size. The size pattern can easily be explained by the higher heterogeneity of the environment for smaller species; for them the environment is coarse-grained. Southwood (1978) has demonstrated how much space of a beech tree can be utilized by small insects. I admit that the size hypothesis is trivial. However, important issues are the precise form of the size-diversity curve (May 1978; Blackburn & Gaston 1994) and the constraints for low and large body size.

The interesting question is if the relative proportion of the species size classes differ between regional pool and local community. If local communities contain a higher percentage of small species, the filling up of the habitat from the regional pool as a source presumably is still going on. It is assumed for many groups (*e.g.* spiders) that smaller species have a higher probability to arrive in the forest and to build up a permanent population. However, in some cases larger, more mobile species (*e.g.* earthworms, some coleopteran groups) may have a better chance to arrive early in the forest during the colonization process.

Hypothesis 2: Diversity is higher for taxa with short-lived species

This assumption is closely linked with hypothesis 2. Lower body size of adults is associated with shorter generation time. It goes without saying that populations with short-lived individuals face a high temporal heterogeneity of their environment. More species can occupy the niche dimension "time". The minority of species in both forests is long-lived (*e.g.* Chilopoda, Diplopoda). Most insects are annual species.

An argument similar to that for low body size postulates that there are constraints for the length of a life-cycle at the lower and the upper limit. Hence the minority of species is long-lived. Hypothesis 2 is not trivial only in the possible case that a higher proportion of species in the forest is short-lived compared to the pool. Assuming that the forest environment is favourable for the species independent of their life-cycle (which is not necessarily the case), colonization would still go on and short-lived species would establish themselves more easily than long-lived ones. However, a reverse trend might be possible if the forest environment favours long-lived species, as compared with open habitats.

Hypothesis 3: Diversity is higher for taxa containing more mobile species

This hypothesis is difficult to test. Similar functional groups with different degrees of mobility should be compared (Lawton 1990), *e.g.* litter spiders and gamasid mites or macropterous and brachypterous beetles. Mobility would be important for the composition of the forest animal community if the colonization process were still going on. However, we have no methodological instrument to decide if species in the pool with low mobility have not yet had the chance to arrive in the forests.

Hypothesis 4: Diversity is higher for taxa with more euryoecious species

This hypothesis would apply to the assumption that more generalist species have a higher chance to reach the forests via habitat stepping stones and to establish themselves in the forests than specialists. As for hypothesis 3, a further premiss should be that a higher percentage of species with more special habitat requirements has not yet had the opportunity to arrive in the forests. Otherwise the relative proportion of generalists and specialists in the pool would be found in the forests, too. The hypothesis might be tested by looking for euryoecy in selected functional groups, *e.g.* earthworms, which is not an easy job. Interestingly, Ulrich (1987) has documented for the hymenopteran parasitoids of the forest "Göttinger Wald" that the ratio parasite species to host species is higher in the forest than in the pool. This means that a relatively higher proportion of generalists occurs in the forest. The almost linear relationship between sampling effort and number of hymenopteran species might be an indication for a continuously high immigration rate and an unsaturated parasitoid community. However, the characteristics of the utilizable host spectrum in the forest is another possible explanation for the high incidence of generalist parasitoids.

Hypothesis 5: Diversity is higher for taxa containing a high proportion of silvicolous species

This hypothesis is a truism. We should expect that taxa with forest species do occur in forests in higher diversities, having the opportunity to utilize the habitat resources and conditions. Many animals groups typical of open habitats are absent from both forests or are species-poor. Examples are syrphid flies, pompilid and sphecid wasps, ants or orthopterans.

Hypothesis 6: Diversity is high for taxa with species utilizing a high diversity of food resources

This point has been discussed above. For the following analysis, I differentiate between five trophic guilds - phytophages (feeders on living plant material), saprophages (utilizers of dead organic matter), microphytophages (feeders on bacteria and/or fungal hyphae and/or algae) and zoophages as predators and parasitoids. According to Schaefer (1996a, 1996b), the species numbers of the five trophic guilds for the taxa Gastropoda, Annelida and Arthropoda in the beech forest "Göttinger Wald" are: 236 phytophages (13.9% of the 1,700 species), 204 saprophages (12.0%), 176 microphytophages (10.4%), 405 predators (23.8%) and 679 parasitoids (39.9%). If the microfauna of the forest is included, the relative proportion of microphytophages is increased: 254 phytophages (13.2% of the 1,918 species), 272 saprophages (14.2%), 270 microphytophages (14.1%), 440 predators (22.9%) and 682 parasitoids (35.6%) (cf. Table 4).

According to a rough calculation, the Central European species pool (of the taxa listed in Table 4) contains 9,900 phytophages (28.3% of the total fauna of about 35,000 species), 4,900 saprophages (13.9%), 3,800 microphytophages (11.0%), 8,000 predators (22.8%) and 8,400 parasitoids (24.0%).

The predictions presented in the discussion on life-forms have been found to apply. Only 62 plant species are the food resource of the species-poor plant-feeders. Saprophagous and microphytophagous populations can utilize diverse food resources in the soil and litter stratum. Interestingly, the fauna based on microflora as food is not more diverse than the detritus-feeding guild. Zoophages meet a high number of populations they can feed upon. Parasitoids are more specialized than predators and surpass this trophic category considerably in species number (cf. discussion in Schaefer 1996a, 1996b).

In comparison with the species pool, phytophages are less represented in the forest "Göttinger Wald". A possible interpretation is that the diversity of the plant-feeding component of the Central European fauna has developed due to specialisation on different habitats (Wiegmann *et al.* 1993; Schaefer 1996a). In contrast, relatively more species of parasitoids are present in the forest than in the pool. The high incidence of generalists in the forest compared to the pool is a possible explanation (Ulrich 1987).

Table 4. Species numbers in the beech forests "Göttinger Wald" ($S_{GÖ}$) and "Solling" (S_{SO}) in comparison with the regional species pool in Central Europe (S_{CE}). % = percentage of the regional species pool occurring in the forests

Animal group	S_{CE}	%	$S_{GÖ}$	%	S_{SO}
Microfauna					
Testacea	900	7.2	65	5.7	51
Turbellaria	150	2	3	2.0	3
Nematoda	1100	5.9	65	8.2	90
Rotatoria	600	2.2	13	?	?
Tardigrada	60	6.7	4	?	?
Saprophagous and microphytophagous mesofauna					
Enchytraeidae	100	36.0	36	15.0	15
Uropodina	184	6.0	11	2.2	4
Cryptostigmata	447	16.8	75	16.1	72
Collembola	300	16.0	48	16.7	50
Saprophagous macrofauna					
Lumbricidae	70	15.7	11	5.7	4
Gastropoda	400	7.5	30	1.0	4
Isopoda	40	15.0	6	0	0
Diplopoda	130	4.6	6	0.8	1
Elateridae	150	7.3	11	2.7	4
Zoophagous mesofauna					
Gamasina	1000	8.0	80	5.3	53
Phytophagous macrofauna					
Heteroptera	800	2.4	19	1.8	14
Auchenorrhyncha	500	3.6	18	0.2	1
Aphidina	850	>0.9	>8	0.1	1
Chrysomelidae	500	2.4	12	1.8	9
Curculionidae	1000	3.3	34	1.2	12
Lepidoptera	3000	1.8	53	1.3	40
Zoophagous macrofauna					
Araneida	900	11.3	102	10.3	93
Pseudoscorpionida	30	10.0	3	6.7	2
Opilionida	40	20.0	8	10.0	4
Chilopoda	50	20.0	10	14.0	7
Carabidae	500	4.8	24	5.2	26
Staphylinidae	1300	6.5	85	9.0	117
Hymenoptera	10000	7.0	704	?	?
Taxa with mixed nutritional biology					
Coleoptera (total)	8000	3.2	254	2.8	225
Diptera	8000	3.7	299	?	?

Hypothesis 7: Local diversity is positively correlated with regional di

The values differ from taxon to taxon. The percentage of species occurr
in relation to the pool in Central Europe ranges from 1.8% (Lepid
(Enchytraeidae) for the forest "Göttinger Wald" and from 0.8% (Diplo
(Cryptostigmata) for the Solling forest (Table 4).

Hypothesis 7 should be tested for similar guilds. Examples are: earthworms - enchytraeids - isopods - diplopods; oribatid mites - collembolans - uropodid mites; spiders - chilopods - carabid beetles - staphylinid beetles - parasitoid hymenopterans; leaf hoppers and plant hoppers (Auchenorrhyncha) - bugs (Heteroptera) - chrysomelid beetles - curculionid beetles. In spite of a considerable variation in colonization percentages, the relationships between species numbers in the regional and the local pool often are in the same order of magnitude for similar guilds of the fauna (cf. the values in Table 4). A prominent case are the 16 to 17% of species of oribatid (= cryptostigmatid) mites and collembolans found in both forests.

Generally, plant-feeders occur in lower percentages (see discussion above). Some groups, such as grasshoppers or tenthredinids, are almost totally absent. Interestingly, only a low number of microfaunal species of the regional pool is found in the forest. The explanation is simple: many of the pool species are strictly aquatic and do not colonize soils. With the exception of the saprophagous macrofauna, both forests exhibit similar tendencies in their colonization patterns.

A positive correlation between local and regional species pool has been found by many authors (*e.g.* Ricklefs 1987; Lawton 1990; Rosenzweig 1995). There is one problem in estimating the local pool: some species are only transient, they cannot establish permanent populations in the forest, they are not "self-maintaining" ("mass effect", Shmida & Wilson 1985).

Hypothesis 8: Diversity is higher for taxa with sources of colonization nearby

This hypothesis is not testable, because the precise macrodistribution of the species of the regional pool cannot be documented for such a high number of populations. Additionally, this hypothesis would only apply under the assumption that not all species have yet arrived in the forest and that the forest community is not saturated.

Hypothesis 9: Communities are more diverse in habitats prone to external forces favouring dispersal

Hypothesis 9 is akin to hypothesis 8. Probably, it cannot be tested for forest habitats.

Hypothesis 10: Communities are more diverse in older habitats

According to the time hypothesis older forests would have caught more species from the regional pool. Such a condition would apply if the community were not saturated and if the colonization process took a long time, longer than the age of the forests. It has been observed that the fauna of old virgin forests is very diverse (Hanski & Hammond 1995), however, naturalness and lack of forest management are parallelled by high spatial heterogeneity, a situation leading to high animal species diversity independent of the colonization time (see hypothesis 13).

The beech wood "Solling" is older than the forest "Göttinger Wald" which was planted in the last century in an open agrarian landscape. However, possible tendencies of colonization patterns are concealed by the structural differences between the two forests.

Hypothesis 11: Communities are more diverse in larger habitats

The theory of island biogeography (MacArthur & Wilson 1967) predicts that larger habitats contain more species (Ricklefs & Schluter 1993; Rosenzweig 1995). This pattern has been observed for many forest animal guilds, *e.g.* bats and birds (Gerell 1988). However, above a certain threshold larger forest patches may be "continents" for invertebrates, the size of the forest is no longer important. In accordance with this expectation, Terrell-Nield (1990) observed that the species number of forest carabids was independent of forest size.

Both forest studied by us are of larger size and patterns of insularity could not be analyzed.

Hypothesis 12: Communities are more diverse in less extreme habitats

It is an undisputed fact that habitats characterized by more extreme environmental conditions are populated by less species ("Thienemann's rule"; Tischler 1993). I do not want to enlarge upon this self-evident pattern.

Basically, the most favourable conditions are met in the ground zone (litter and upper soil layers, herb layer). The higher vegetation layers are dominated by adverse factors such as low moisture, high temperatures, secondary plant substances in food; with increasing soil depth the availability of oxygen and of food resources decreases, the levels of carbon dioxide rise. Consequently, forest species diversity reaches its peak in the layers of the ground zone.

In comparison with the mull-structured forest "Göttinger Wald", the mor environment in the "Solling" is more extreme: adverse factors are acidity of soil, low organic content in the upper soil layer, lack of herbs. It is tempting to explain the lower species diversity in the latter habitat at least partly by the adversity of the environment.

Hypothesis 13: Communities are more diverse in spatially heterogeneous habitats

The parallelism between structural diversity and species diversity is well-established. The complex spatial structure of the soil is supposed to be one determinant of high species diversity of soil animals (Anderson 1975). However, Hutchinson (1961) has emphasized that a homogeneous spatial matrix does not exclude the development of a highly diverse fauna and flora; the upper pelagic zone is inhabited by a large number of plankton populations.

In the forest "Göttinger Wald" two components of high spatial diversity can be differentiated. The herbs form a complex vegetation layer, favouring populations of the interface litter - herb layer, *e.g.* some spider species. The bioturbating macrofauna (mainly earthworms) creates a complex mixture of microhabitats in the upper centimetres of the soil, utilizable by a number of populations of the mesofauna and microfauna.

Many studies confirm the positive relationship between spatial heterogeneity and animal species numbers. An important component are plant species richness and vegetational diversity (Andow 1991; Huston 1994). Jones & Lawton (1991) explain the relationship between umbellifers and insects with chemical heterogeneity of the plant species. Gaston (1992) found a positive correlation between the number of plant species and insect species; however, the increase in insect species diversity was lower than the increase in plant species diversity. Hanski & Hammond (1995) observed higher species numbers in spatially more complex forests; contrary to this, Gerrell (1988) detected no clear relationship between forest vegetation structure and number of invertebrates.

Hypothesis 14: Communities are more diverse in temporally heterogeneous habitats

Temporal heterogeneity is a correlate to spatial heterogeneity. For populations with a short life-cycle the environment is coarse-grained, and it can be expected that habitats with a complex phenology offer niches for a higher number of species. A test of this plausible hypothesis is difficult and not possible for the two studied forests.

Hypothesis 15: Communities are more diverse in habitats with more diverse food resources

Habitats with a higher diversity of certain categories of food (*e.g.* plant species, litter species, microflora) should support a higher number of animal species. This assumption is a complement to hypothesis 6 which focusses on the life form of the species.

Are there differences in trophic organisation between the mull-structured and the mor-structured forest? Some patterns emerge: In the mor forest, phytophages occur in lesser numbers, obviously because a herb layer is not present and plant species diversity is low. Microphytophages are comparable in species diversity to the mull-structured forest, obviously because the mor soil and litter are populated by a high number of fungal populations (however, precise data are lacking). Macrosaprophages are almost absent from the adverse acid mor environment; microsaprophages occur in higher numbers, comparable to the mull-structured forest. Predator diversity does not remarkably differ between both forests; obvious factors are diversity of prey populations and the structural diversity of the litter layer. For parasitoids no data are available for the Solling beech forest.

Hypothesis 16: Communities are more diverse in habitats where biotic factors (such as enemy pressure or interspecific competition) are prevalent

The role of predation and interspecific competition for community structure is emphasized in every textbook of ecology (*e.g.* Ricklefs 1990; Putman 1994); it is generally assumed that high predation rates and high incidence of interspecific competition leads to diversification of food webs and species numbers (Menge & Sutherland 1976). Grazing by enemies may enhance diversity in the assembly of prey populations (*e.g.* Paine 1994). Competition may lead to niche partitioning (*e.g.* Lawton 1990; Rosenzweig 1995).

Comparing forests with different degrees of biotic interactions is a difficult job for two reasons. It is not easy to quantify biotic interference within the forest animal community. A further hurdle is the proof of a biotic effect on animal population presence or absence, this is especially true for indirect mutualistic relationships. Are collembolan species numbers influenced by the high degree of predation? Do parasitoids influence species diversity of herb and canopy phytophages? No answer is possible. It is generally assumed (but has not been demonstrated experimentally in the field) that microfloral diversity is partly the consequence of grazing by microphytophages. The role of past or present competition for the organisation of forest animal communities is a merely speculative issue. For instance, Müller (1985) explains the diversity of forest carabid communities with historic and present competitive interactions in different niche dimensions without any convincing descriptive analysis or experimental proof.

Hypothesis 17: Communities are more diverse in more productive habitats

The production hypothesis is discussed in nearly all publications about diversity (Putman 1994; Rosenzweig 1995). Many descriptive studies led to proof or disproof of this assumption.

Both forest studied have equal rates of production, but differ in their species diversity. Hence they cannot be used as an indication for the production hypothesis.

Some further studies are given as examples. Currie (1991) found that more productive habitats had more animal species. Latham & Ricklefs (1993) could not detect a relationship between forest productivity and number of tree species. In water-filled containers species numbers were lower in less productive habitat units (Jenkins *et al.* 1992).

Hypothesis 18: Communities are more diverse in habitats with a high stochastic variation of environmental factors

Variation in the environment may favour higher species numbers; on the other hand stochastic variability of conditions for the development of populations can be adverse.

Hypothesis 19: Communities are more diverse in disturbed habitats

At small-scale intermediate levels of disturbance species diversity tends to increase (Pickett & White 1985; Rosenzweig 1995). This is the message of the "intermediate disturbance hypothesis". Numerous studies have confirmed the disturbance-diversity pattern.

It is tempting to interpret the higher species numbers in the mull-structured forest with the disturbance caused by bioturbation macrofauna populations, above all earthworms. They transport organic material into the upper layer of the mineral soil and change the soil environment by their burrowing activity. A patchwork of habitats on the microscale and the mesoscale may favour the establishment of many populations of the microfauna and mesofauna.

Conclusions

My purpose was by no means to test a set of hypotheses and reach a definite answer. I wanted to open the reader's eyes for the possible structure of hypotheses for future studies. In summary, not much is known about factors determining species diversity in a forest. Some plausible patterns exist, but the many possible ecological or evolutionary factors are interrelated in determining observed diversity patterns in forest animal communities.

Many of the hypotheses are difficult to evaluate; some of them are common-sense insights. To my mind it became evident that regional species diversity, trophic type of the taxon in question, food resources, disturbance regime and heterogeneity in the habitat are factors of special importance.

The precise knowledge of cause and effect in diversity patterns would help to tackle the many unsolved questions in the fields diversity-stability, function of diversity and diversity in food webs.

References

Alphei, J. 1994. Die freilebenden Nematoden von Buchenwäldern mit unterschiedlicher Humusform: Struktur der Gemeinschaften und Funktion in der Rhizosphäre der Krautvegetation. Berichte des Forschungszentrums Waldökosysteme Göttingen A 125: 1-165.

Anderson, J.M. 1975. The enigma of soil animal species diversity. pp. 51-58. In: Vanek, J. (ed), Progress in Soil Zoology. Academia, Prague. Junk, The Hague.

Andow, D.A. 1991. Vegetational diversity and arthropod population response. Annual Review of Entomology 36: 561-568.

Blackburn, T.M. & Gaston, K.J. 1994. Animal body size distributions: patterns, mechanisms and implications. Trends in Ecology and Evolution 9: 471-474.

Cornell, H.V. 1993. Unsaturated patterns in species assemblages: The role of regional processes in setting local species richness. pp. 243-252. In: Ricklefs, R.E. & Schluter, D. (eds), Species Diversity in Ecological Communities. University of Chicago Press, Chicago, London.

Cornell, H.V. & Lawton, J.H. 1992. Species interactions, local and regional processes, and limits to the richness of ecological communities: a theoretical perspective. Journal of Animal Ecology 61: 1-12.

Currie, D.J. 1991. Energy and large scale patterns of animal and plant species richness. American Naturalist 137: 27-49.

Dunger, W. 1983. Tiere im Boden. 3rd ed. Die Neue Brehm-Bücherei 327. Ziemsen, Wittenberg.
Ellenberg, H., Mayer R. & Schauermann, J. 1986. Ökosystemforschung. Ergebnisse des Sollingprojekts 1966-1986. Ulmer, Stuttgart.
Eriksson, O. (1993): The species-pool hypothesis and plant community diversity. Oikos 68: 371-374.
Fitton, M.G., de V. Graham, M.W.R., Boucek, Z.R.J., Fergusson, N.D.M., Huddleston, T., Quinlan, J. & Richards, O.W. 1978. A Check List of British insects. Part 4: Hymenoptera. Handbooks for the Identification of British Insects, Vol. XI, Part 4. Royal Entomological Society of London.
Gaston, K.J. 1992. Regional numbers of insects and plant species. Functional Ecology 6: 243-247.
Gerell, R. 1988. Faunal diversity and vegetation structure of some deciduous forests in south Sweden. Holarctic Ecology 11: 87-95.
Günther, H. & Schuster, G. 1990. Verzeichnis der Wanzen Mitteleuropas. Deutsche Entomologische Zeitschrift N.F. 37: 361-396.
Hannemann, H.-J., Klausnitzer, B. & Senglaub, K. 1992. Exkursionsfauna von Deutschland. Volume 1. Wirbellose (ohne Insekten). 8th ed. Gustav Fischer, Jena, Stuttgart.
Hannemann, H.-J., Klausnitzer, B. & Senglaub, K. 1994. Exkursionsfauna von Deutschland. Volume 2. Wirbellose: Insekten. 8th ed. (volume 2/1). 7th ed. (volume 2/2). Gustav Fischer, Jena, Stuttgart.
Hanski, I. & Hammond, P. 1995. Biodiversity in boreal forests. Trends in Ecology and Evolution 10: 5-6.
Heimer, S. & Nentwig, W. 1991. Spinnen Mitteleuropas. Paul Parey, Berlin, Hamburg.
Hövemeyer, K. 1992. Die Dipterengemeinschaft eines Kalkbuchenwaldes: eine siebenjährige Untersuchung. Zoologische Jahrbücher Systematik, Ökologie und Geographie 119: 225-260.
Hövemeyer, K. 1995. Die Dipterengemeinschaften einiger terrestrischer Ökosysteme in Südniedersachsen: eine ökologische Analyse. Habilitationsschrift, Göttingen.
Huston, M.A. 1994. Biological Diversity. Cambridge University Press, Cambridge.
Hutchinson, G.E. 1961. The paradox of plankton. American Naturalist 95: 137-145.
Jacobs, W. & Renner, M. 1988. Biologie und Ökologie der Insekten. 2nd ed. Gustav Fischer, Stuttgart.
Jenkins, B., Kitching, R.L. & Pimm, S.L. 1992. Productivity, disturbance and food web structure at a local spatial scale in experimental container habitats. Oikos 65: 249-255.
Jones, C.G. & Lawton, J.H. 1991. Plant chemistry and insect species richness of British umbellifers. Journal of Animal Ecology 60: 767-777.
Jones, C.G., Lawton, J.H. & Shachak, M. 1994. Organisms as ecosystem engineers. Oikos 69: 373-386.
Karg, W. 1989. Acari (Acarina), Milben - Unterordnung Parasitiformes (Anactinochaeta) - Uropodina Kramer, Schildkrötmilben. Tierwelt Deutschlands, Teil 67. Gustav Fischer, Jena.
Karg, W. 1993. Acari (Acarina), Milben - Unterordnung Parasitiformes (Anactinochaeta) - Cohors Gamasina Leach, Raubmilben. 2nd ed. Tierwelt Deutschlands, Teil 59. Gustav Fischer, Jena.
Kerney, M., Cameron, R. & Jungbluth, J. 1983. Die Landschnecken Nord- und Mitteleuropas. Paul Parey, Hamburg, Berlin.
Kloet, G.S. & Hincks, W.D. 1976. A check list of British insects. Part 5: Diptera and Siphonaptera. Handbooks for the Identification of British Insects, Vol. XI, Part 5. Royal Entomological Society of London.
Kloet, G.S., Hincks, W.D. & Pope, R.D. 1977. A check list of British insects. Part 3: Coleoptera and Strepsiptera. Handbooks for the Identification of British Insects, Vol. XI, Part 3. Royal Entomological Society of London.
Krebs, C.J. 1989. Ecological Methodology. Harper & Row, New York, Cambridge.
Latham , R.E. & Ricklefs, R.E. 1993. Global patterns of tree species richness in moist forests: energy-diversity theory does not account for variation in species richness. Oikos 67: 325-333.
Lawton, J.H. 1990. Species richness and population dynamics of animal assemblages. Patterns in body size: abundance space. Philosophical Transactions of the Royal Society of London B 330: 283-291.
Lawton, J.H. 1994. What do species do in ecosystems? Oikos 71: 367-374.
Lawton, J.H. & Brown, V.K. 1993. Redundancy in ecosystems. pp. 255-270. In: Schulze, E.-D. & Mooney, H.A. (eds), Biodiversity and Ecosystem Function. Ecological Studies 99. Springer, Berlin, Heidelberg, New York.
Lucht, W. 1987. Die Käfer Mitteleuropas. Catalogue. Goecke & Evers, Krefeld.
Mac Arthur, R.H. & Wilson, E.O. 1967. The Theory of Island Biogeography. Princeton University Press, Princeton.
Martens, J. 1978. Weberknechte, Opiliones. Tierwelt Deutschlands, Teil 64. Gustav Fischer, Jena.
May, R.M. 1978. Dynamics and diversity of insect faunas. pp. 188-204. In: Mound, L.A. & Waloff, N. (eds), Diversity of Insect Fauna. Blackwell Scientific, Oxford.

Menge, B.A. & Sutherland, J.P. 1976. Species diversity gradients: Synthesis of the roles of predation, competition, and temporal heterogeneity. American Naturalist 110: 251-369.
Müller, J.K. 1985. Konkurrenzvermeidung und Einnischung bei Carabiden (Coleoptera). Zeitschrift für zoologische Systematik und Evolutionsforschung 23: 299-314.
Paine, R.T. 1994. Marine Rocky Shores and Community Ecology: An Experimentalist's Perspective. Excellence in Ecology 4. Ecology Institute, Oldendorf.
Pickett, S.T.A. & White, P.S. 1985. The Ecology of Natural Disturbance and Patch Dynamics. Academic Press, New York.
Putman, R.J. 1994. Community Ecology. Chapman & Hall, London, Glasgow.
Ricklefs, R.E. 1987. Community diversity: Relative roles of local and regional processes. Science 235: 167-171.
Ricklefs, R.E. 1990. Ecology. 3rd ed. Freeman, New York.
Ricklefs, R.E. & Schluter, D. (eds). 1993. Species Diversity in Ecological Communities. University of Chicago Press, Chicago, London.
Rosenzweig, M.L. 1995. Species Diversity in Space and Time. Cambridge University Press, Cambridge.
Schaefer, M. 1990. The soil fauna of a beech forest on limestone: trophic structure and energy budget. Oecologia 82: 128-136.
Schaefer, M. 1991. Animals in European temperate deciduous forests. pp. 503-525. In: Röhrig, E. & Ulrich, B. (eds), Temperate Deciduous Forests (Ecosystems of the World). Elsevier, Amsterdam.
Schaefer, M. 1994. Brohmer - Fauna von Deutschland. 19th ed. Quelle & Meyer, Heidelberg, Wiesbaden.
Schaefer, M. 1996a. Die Artenzahl von Waldinsekten: Muster und mögliche Ursachen der Diversität. Mitteilungen der deutschen Gesellschaft für allgemeine und angewandte Entomologie. In press.
Schaefer, M. 1996b. Diversity of the animal community in a beech forest - patterns and processes. Proceedings of the 6th International Conference of the Israel Society for Ecology and Environmental Quality Sciences, "Preservation of Our World in the Wake of Change", Jerusalem, 1996. In press.
Schaefer, M. & Schauermann, J. 1990. The soil fauna of beech forests: comparison between a mull and a moder soil. Pedobiologia 34: 299-314.
Schulz, E. 1991. Die Milbenfauna (Acari: Mesostigmata und Cryptostigmata) in Lebensräumen auf Kalkgestein: Populationsökologie, Sukzession und Beziehungen zum Lebensraum. Dissertation, Göttingen.
Shmida, A. & Wilson, M.V. 1985. Biological determinants of species diversity. Journal of Biogeography 12: 1-20.
Smith, F.D.M., May, R.M., Pellew, R., Johnson, T.H. & Walter, K.R. 1993. How much do we know about the current extinction rate? Trends in Ecology and Evolution 8: 375-378.
Southwood, T.R.E. 1978. The components of diversity. pp. 19-40. In: Mound, L.A. & Waloff, N. (eds), Diversity of Insect Fauna. Blackwell Scientific, Oxford.
Terrell-Nield, C. 1990. Is it possible to age woodlands on the basis of their carabid beetle diversity? Entomologist 109: 136-145.
Tischler, W. 1993. Einführung in die Ökologie. 4th ed. Gustav Fischer, Stuttgart, Jena.
Ulrich, W. 1987. Wirtsbeziehungen der parasitoiden Hautflügler in einem Kalkbuchenwald (Hymenoptera). Zoologische Jahrbücher, Abteilung für Systematik, Ökologie und Geographie der Tiere 114: 303-342.
Walker, B.H. 1992. Biodiversity and ecological redundancy. Conservation Biology 6: 8-23.
Weigmann, G. 1981. Die deutschen Hornmilbenarten und ihre ökologische Charakteristik. Zoologische Beiträge N.F. 27: 459-489.
Wiegmann, B.M., Mitter, C. & Farrell, B. 1993. Diversification of carnivorous insects: extraordinary radiation or specialized dead end? American Naturalist 142: 737-754.
Wilcke, D.E. 1963. Oligochaeta. Tierwelt Mitteleuropas, Volume 1, Lief. 7a. Quelle & Meyer, Leipzig.
Wilson, E.O. 1992. The Diversity of Life. Harvard University Press, Cambridge.

CHAPTER 2
FAUNA, FLORA, AND VEGETATION IN ECOSYSTEMS: SOME ASPECTS OF BIODIVERSITY

BIODIVERSITY IN MEDITERRANEAN ECOSYSTEMS

GIUSEPPE PIGNATTI & SANDRO PIGNATTI

Dipartimento di Biologia Vegetale, Univ. di Roma "La Sapienza", Città Universitaria 00185 Roma, Italy

Key words: coevolution, diversity, evergreen vegetation, Mediterranean flora

Abstract

The flora of the Mediterranean basin includes ca. 20,000 species of vascular plants and can consequently be regarded as one of the richest on the globe. The highest diversity is in general found in the Northern portion and on mountains, and appears concentrated in the belt with an optimal combination of warm temperatures and high rainfall. The topographical distribution of species is not uniform: climax vegetation has generally a feebly diversified flora; endemics mostly occur in extreme habitats (coasts, mountains), whereas the richest flora grows in environments influenced by man. The hypothesis of man-vegetation interactions in the garrigue is discussed.

The concept of biodiversity

Biodiversity deals with the result of evolution, one of the basic processes of life. Evolution consists in differentiation and selection of genotypes, of course, biodiversity cannot be measured only by the genetic distance between genotypes. Let us compare two couples of woody species: *Acer obtusatum* - *A. opalus* and *Pinus mugo* - *P. uncinata*. In the first case the only difference lies in the distribution of hairs on leaves, whereas in the second case the most impressive differences are in habitus and ecology (*P. mugo* has creeping stems and forms a compact heath, *P. uncinata* is an erect tree living in forests), and the small differences in morphological characters are not very significant. The genetic distance of the two couples is probably very small (both include closely related species and are probably monophyletic), but the process of diversification

A. Kratochwil (ed.), Biodiversity in ecosystems, 59-73

appears much wider in *Pinus* than in *Acer*. In fact, in our opinion biodiversity can be considered as the expression of biological evolution focused on differences in morphology, physiology and ecology. The Mediterranean flora developed in a highly diversified environment and was strongly influenced by geological and climatic changes of the past, and more recently by the action of man. The plant life of this region can be discussed as an example of biodiversity in ecosystems.

The Mediterranean basin

The Mediterranean zone consists of the Mediterranean sea, countless islands (large and small ones) spread over the sea, some major peninsulas and the surrounding regions; this zone is not limited by a clear geographical boundary and is characterized chiefly by a unitary climate. The total land surface can be estimated to amount to approximately 1,500,000 sq.km., to which ca. 3 million sq.km. of inland sea should be added.

The climate of the Mediterranean zone is a warm temperate one (yearly average temperatures: 14-18 °C) with an annual rainfall of ca. 400-1,000 mm; the rainfall is concentrated in the cold season, whereas summers are almost completely dry. The summer drought is the most relevant feature of Mediterranean-type climates (which occur in 4 other regions of the world: California, Chile, Western and Southern Australia, and South Africa); in this sense, the climate of the Mediterranean zone is very different from other warm temperate or subtropical climates (*e.g.* China, Japan, Eastern U.S.A., Argentina, Queensland etc.) which are characterized by dry winters and summer rain. Temperatures are relatively mild: in winter no monthly average below 0 °C, frost days and snowfall are exceptional; on summer days maxima over 40 °C are also exceptional. Because of the combination of high temperatures and drought, summer is a period of severe water stress for plants living in the Mediterranean area: in general the dry period lasts 3 months, sometimes longer (in the Eastern and Southern parts up to 5-6 months) or shorter (at the Northern boundary).

The flora of the Mediterranean zone is estimated at ca. 20,000 species (Greuter *et al.* 1984-1989), consequently it can be considered a very rich one. In comparison, the flora of the whole European continent (with a surface 6 times larger), comprises "only" 10,000 species (Webb 1984), *i.e.* the species density in the Mediterranean area is 12 times higher than that of Europe. In addition, the species total of the Mediterranean flora is not inflated by groups with swarms of agamospecies: the only group with a large number (ca. 200) of apomicts is *Limonium*, whereas *Hieracium, Taraxacum, Alchemilla* and *Rubus* are little represented in the Mediterranean flora.

For most botanists from Central European countries, the Mediterranean zone is the nearest area with elevated diversity, and many students come from the North to investigate its flora (*e.g.* the Englishman Sibthorp and later the Austrian Halacsy for the Greek flora, the German Willkomm for the Spanish flora, Davis for Turkey, Rikli from Zürich for the whole Mediterranean vegetation, and many others). On the contrary, no relevant investigations have been carried out in Central and Northern parts of the continent by botanists coming from Southern Europe.

Floristic diversity

Up to now a general flora of the Mediterranean zone does not exist; only floras for the individual countries or local ones are available, indeed the literature for some regions is rather poor. The only floristic treatments of larger areas are the flora of the Orient by Boissier (1867-1884) and the Balkan flora by Hayek & Markgraf (1924-1933) (both publications are of a high standard, however, meanwhile they have - more or less - merely historical value), and the flora of North Africa begun by Maire (1952-1980) which is still in progress (only ca. 1/3 has been published in a period of over 30 years). Flora Europaea includes only part (ca. 40 %) of the Mediterranean zone. The redaction of a checklist supported by the Organization for Plant Taxonomical Investigation of the Mediterranean Area (OPTIMA) as strategic program MED-CHECK is still in progress (ca. 50 % of the families published, Greuter *et al.* 1984-1989) and will hopefully be concluded in 5-7 years' time. The estimated number of Mediterranean species has been derived from the projection of MED-CHECK figures.

Under these conditions it is quite difficult to give some figures of the species diversity in the Mediterranean zone, any conclusion depending on the data which are respectively used as a basis for estimations.

The number of species in the European countries facing the Mediterranean sea (computations by Webb 1984, based on Flora Europaea) is given in Table 1.

Table 1. Species numbers and surfaces of European countries

country	a species number (Webb 1978)[1]	b surface sq.km.	index a/b
Europe	11,047	10,107,000	1.09
Italy	5,300	251,479[2]	21.1
Spain	5,200	493,515[3]	10.5
Yugoslavia	5,075	256,393[5]	19.8
France	4,375	535,285[4]	8.2
Germany	2,675	353,640[5]	7.5
Poland	2,350	311,73	7.5
British Isl.	1,775	244,768	7.2
Norway	1,500	323,917	4.6
Finland	1,350	337,009	4.0
Sweden	1,700	449,531	3.8

[1] - average. [2] - without Sardinia and Sicily. [3] - only continental territory. [4] - without Corsica. [5] - surfaces in 1978.

It should be observed that in general the species concept used in Flora Europaea is somewhat wide, and many local endemics, chiefly from Mediterranean mountain areas, are not treated as separate taxa; consequently the number of indigenous species may increase by some percent: our flora of Italy describes over 5,600 species (indigenous or frequently naturalized), including Sicily, Sardinia and Corsica (with ± 200 endemics) and the flora of Spain may increase to 6,000 species (Rivas-Marínez, verbal comm.).

The data of Webb (1984) show two evidences:

a) Floristic diversity increases from North to South, reaches a maximum apparently at the boundary between the Central European and North Mediterranean zone and then again declines in the South (Fig. 1).

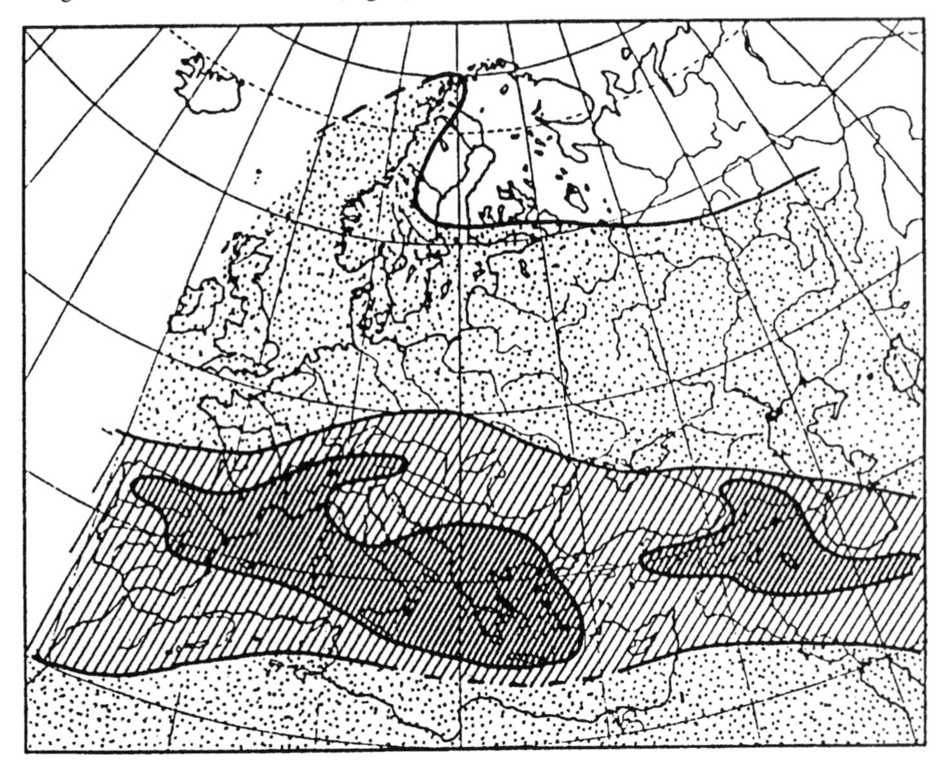

Figure 1. Floristic density in Europe and the Mediterranean basin. The intervals are: more than 3,000 species on 100,000 sq.km., 2,000-3,000 species, 1,000-2,000 species, less than 1,000 species.

The number of occurring species seems to depend on temperature and water availability; temperature is minimal in the North and increases regularly toward, the South; as to water, rainfall alone is not significant and has to be related to evaporation: in this sense, the quantity of water available for plant growth is maximal in Northern and Central Europe, and decreases rapidly in the Mediterranean zone. Consequently the boundary between Central Europe and the Mediterranean zone is where both factors occur in the most convenient combination.

b) A direct relationship seems to exist between the area of each country and the number of species occurring, but floristic diversity also depends on the environmental character of the different territories (Fig. 2): countries with diversified topography (*e.g.* Greece, Italy, Spain) have higher figures than the relatively uniform ones (*e.g.* Finland, Germany, Poland).

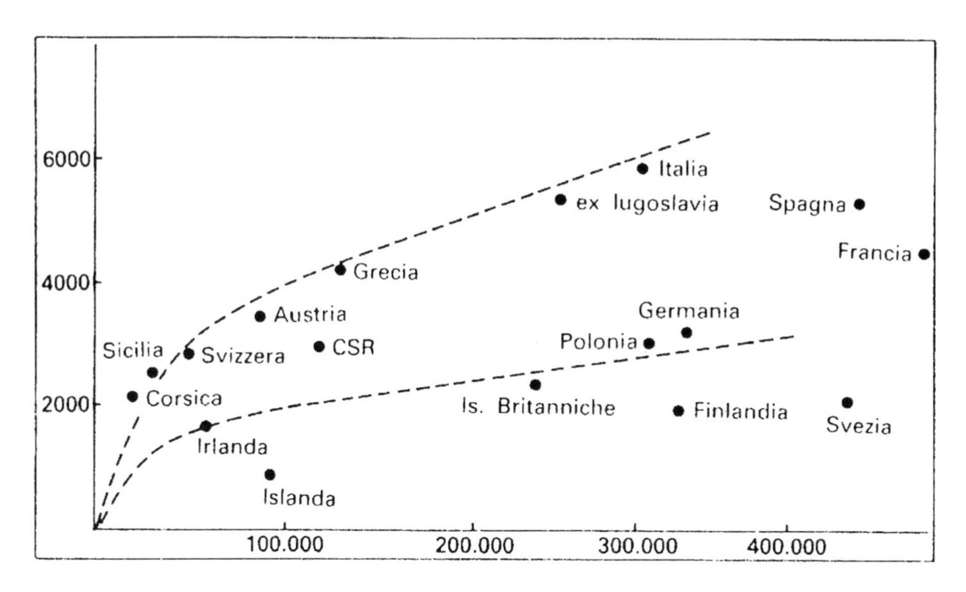

Figure 2. Relationships between number of species and surface. Upper line: countries with diversified topography; lower line: uniform territories.

Sicily and Malta have about the same climate and phytogeographical connections, but Sicily is larger and the number of species occurring there is higher by a factor of ca. 2.5. In fact, there is probably not only a correlation with surface, but also - and chiefly - with niche-diversity: a larger country evidently has more diversified niches than a small one. It may be pointed out that Malta possesses only low hills (max. altitude 258 m), whereas Sicily has mountains reaching ca. 2,000 m, as well as the Etna volcano with ca. 3,300 m altitude, and in consequence comprises a lot of mountain species which are completely lacking in Malta.

A more comparable estimation of the floristic density can be based on objective procedures of floristic inventory, as carried out by Perring & Walters (1962), Haeupler & Schoenfelder (1988), and for NE Italy by Poldini (1991). A large area in Central Italy was investigated. Apparently the Mediterranean zone cannot be considered as being particularly rich in species (Table 2): concentrations of 1,000 species and more, which can be found at the Southern boundary of the Alps (Pre-Alps) are definitely richer, probably because of the variety of moist niches (streams, creeks, swamps and mires), which are rare or almost inexistent in the Mediterranean mountains. Consequently it must be assumed that a special interest in the Mediterranean flora is not due to the quantity of species alone, but to some qualitative features (richness in endemics or disjunct species, plants with ancestral characters, connections with exotic floras etc.).

Table 2. Species numbers on quadrants of 35 sq.km.

	Alps	Mediterranean
mountains		
with continental climate	450-600	-
with suboceanic climate	500-800	450-800
prealpine	750-1,000	450-800
hills	350-500	300-450
cultivated plain	200-250	250-300
coasts	300-400	300-400
investigated quadrants (tot.)	160	91

Evolution of the Mediterranean flora

Some events which marked paleogeography of the Mediterranean and some evolutionary processes may be shortly summarized (for more details cf. Pignatti 1995):
a) the original dualism between the Western component (**Laurus,** Crassulaceae, **Cistus,** Genisteae, **Arbutus, Erica, Micromeria, Senecio**) and the Eastern one (Fagaceae, Chenopodiaceae, **Astragalus**, Primulaceae);
b) expansion of the Eastern component during Messinian;
c) expansion of the evergreen forest vegetation (**Pinus, Abies**, the **Ilex-Taxus** belt) on mountains at the end of Pliocene;
d) plate tectonics and connections of the West-Mediterranean flora to Corsica-Sardinia and of the East-Mediterranean flora to Apulia and Gargano;
e) expansion of the deciduous forest after glaciations.

These features form the general context for the process of evolution through reduction, which takes place under the influence of man and as a form of the coevolution plant-man. This process is characteristic for the Mediterranean and is lacking in other zones with a climate of Mediterranean type on the other continents. It accounts for the high floristic diversity of this area.

Evolution through reduction

The vegetation of the Mediterranean zone is well-known, and at least for Spain, Southern France, Italy (islands included) and parts of North Africa an exhaustive survey of the occurring plant associations is possible. Hundreds of vegetation types have been described and the frequency of species among them has been studied in many investigations.

Under natural conditions climax vegetation covers large areas and lasts for a long time: consequently it may be supposed that highest diversity should be found in climax associations. This is what happens in the tropical environment. In the Mediterranean zone, on the contrary, diversity appears presently concentrated in the secondary vegetation. This finding needs to be discussed.

The following general features can be recognized:
1. Climax vegetation is represented by the evergreen oak forest (**Quercetum ilicis**) in regions with higher rainfall and the chaparral of evergreen treelets and shrubs (**Oleo-Ceratonion**) in the driest parts (Pignatti 1986). Diversity in this vegetation is very low as regards the number of species present (cf. Table 3).

Table 3. Floristic diversity of the Mediterranean vegetation (from Pignatti 1988)

	Southern France (Braun-Blanquet 1952)			Barcelona (Bolos 1962)		
	genera	species	ass.	genera	species	ass.
Evergreen forest	58	69	2	58	66	2
Macchia	65	78	2	60	68	2
Garriga	235	435	28	171	286	10

The number of associations which can be distinguished is also very low: *e.g.* all forest types with **Quercus ilex** in the whole Mediterranean zone (from Spain to Greece) can be described with no more than 5 associations.
2. Climax vegetation is mostly substituted by secondary maquis or garrigue, consisting of low shrubs, and here the diversity of species present as well as of associations is very high (see Table 3).
3. The qualitative uniqueness of the Mediterranean flora is not focused in the climax vegetation, where endemics are mostly absent, nor in the secondary vegetation which greatly depends on human impact, but appears concentrated in marginal habitats (rocks, mountains, sea cliffs etc.), which are rich in endemic or localized species.

A modern approach to this problem is possible through application of the information theory. Vegetation can be considered as a system exchanging information with the environment. An evaluation of the information flow is possible by employing the Shannon-Weaver formula and can be expressed as entropy values or as neg entropy values, *i.e.* as a measure of order (= low entropy). A serious limitation is given by the fact that phytosociological relevés in general are not suitable for the application of this method, because of the insufficient precision of cover values, so that it is necessary to carry out more exact estimations of frequency. This method gives some new points of view but probably is not sufficient for a general treatment of the problem. The primary succession and the most important factors effective in the Mediterranean vegetation as to grazing and fire can be investigated from this point of view and some results can be discussed.

The primary succession on the Mediterranean coast was investigated by Celesti & Pignatti (1988). Entropy values are regularly increasing from the pioneer stages to the more ripe vegetation, *e.g.* on the coast near Rome (Table 4).

Table 4. Entropy values of vegetation on the coast near Rome

	entropy
Cakiletum	1.48
Agropyretum	1.83
Ammophiletum	1.75
Crucianelletum	1.99
Juniperetum phoeniceae-macrocarpae	1.91
Quercetum ilicis	2.29

The increasing values appear on the one hand to be a consequence of the growing number of species occurring in the various associations, on the other side they seem correlated with the fact that in pioneer vegetation only one species dominates, whereas in more ripe stages several species have more or less similar cover values. The transition from the herbaceous dune vegetation to the last two associations, which are composed mainly of woody species, is followed by a sharp decline in entropy values, so that the succession in the herbaceous vegetation and that in the woody one appear as two distinct and relatively independent phenomena. In this sense it seems difficult to conclude that during succession entropy is increasing; it is more likely to give the opposite interpretation, *i.e.* that salinity is acting as a selective factor and in consequence entropy is lower in the pioneer associations occurring on the shore.

The consequences of grazing were investigated by Naveh & Whittaker (1979). In Israel the final vegetation is characterized by a comparatively low level of entropy (Fig. 3); by grazing entropy is progressively increasing, but when the grazing pressure becomes very strong, entropy declines again. A linear correlation between grazing pressure and entropy in vegetation does apparently not exist. In fact, a limited grazing activity causes an increase in entropy, *i.e.* low values of order, and this seems a consequence of the mechanical impact on vegetation: cattle and sheep pick up plant material and produce small gaps in the herb layer, and the same is caused by the hooves; in the gaps new plants, mostly belonging to alien species, can germinate and are revealed by the entropy value as a factor of disorder. Also the dejections of grazing animals, which are rich in nutrients, create new niches that can be colonized by alien species. When grazing pressure becomes stronger, on the contrary, a severe selection occurs, and palatable species are consumed; at the end only poisonous or spiny species remain, and the entropy value declines again. It is a bit difficult to consider this as an increase of order, but really it is so, at least from a mathematical point of view.

The effect of fire on Mediterranean vegetation is generally a condition of disorder, whereas the species composition is not significantly changed, at least when fire remains a more or less sporadic event. Data from the Languedoc (Trabaud & Lepart 1980), from the Thyrrhenian coast (Fig. 4, from De Lillis & Testi 1989) and from the submediterranean vegetation near Trieste (Feoli *et al.* 1981) demonstrate that immediately after fire entropy is increasing. This is a consequence of the more or less complete destruction of subaerial phytomass: competition for space and light is lower and quantitative relationships among species become more even; the situation is probabilistic and can be described as higher entropy. In the case of Trieste, observations have been continued from a fire in 1968 up to the present. During more than 20 years after the fire, the species composition remained unchanged and the normal development of vegetation was characterized only by regression of the majority of species, caused by the progressive expansion of a few, slowly becoming dominants. The process is very slow and 20 years after the fire, the dominance relationships still remain different from the previous situation.

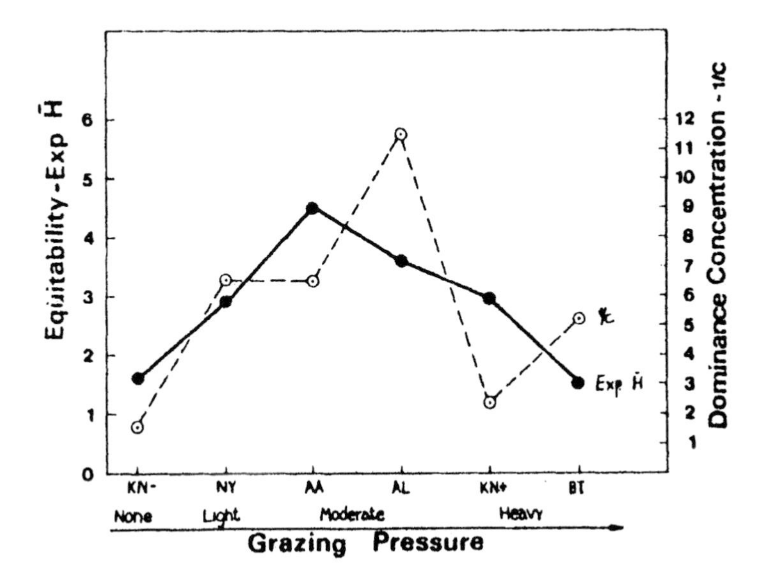

Figure 3. Dominance concentration reciprocal (1/C) and equitability (exp H') of vascular plant species of open woodlands and shrublands in Israel, in relation to grazing pressure. From Naveh & Whittaker 1979.

Only by repeated fire highly specialized vegetation types occur (*Cisto-Lavanduletea, Tuberarietea, Rosmarinetalia*) and remain as more or less permanent features.

In conclusion, the application of the concept of entropy to understand successions in vegetation appears of limited value. In fact, entropy is a concept of reversible thermodynamics, elaborated for the study of linear systems. It can be hardly utilized for vegetation, which is a complex system. Only in such cases when succession can be explained on the basis of a single factor (as for salinity or fire) the system can be linearized at least for a short time and consequently the use of entropy can produce good results.

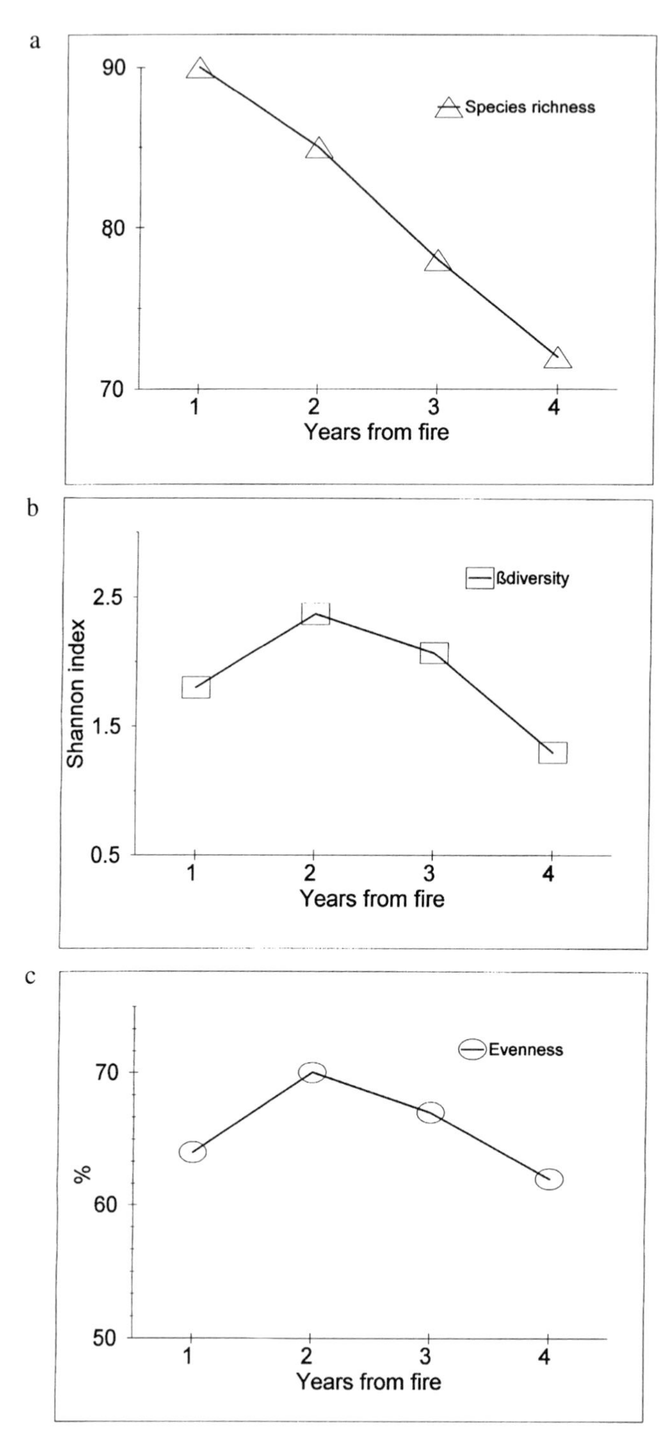

Figure 4. Floristic richness (a), diversity (b) and evenness (c) in the Mediterranean vegetation of C. Italy (*Quercion ilicis*, *Thero-Brachypodion*) over four years after burning. From De Lillis & Testi 1989.

Coevolution plant-man in the Mediterranean area

A quantitative comparison of floristic and vegetation diversity among floras of marginal habitats, climax vegetation, and secondary maquis and garrigue is given in Table 3 and Fig. 5. The relationships are summarized in Table 5.

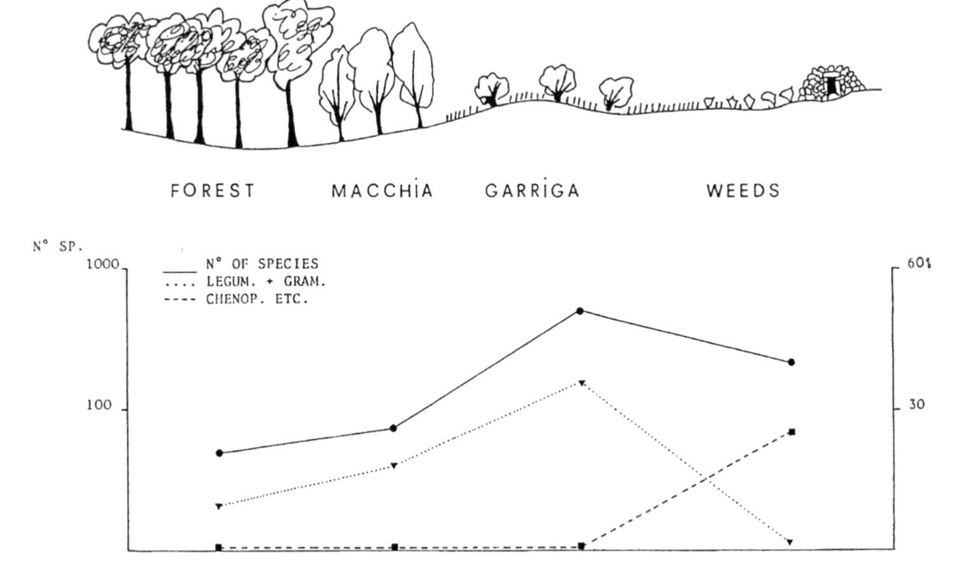

Figure 5. Floristic diversity in different types of Mediterranean vegetation. Number of species present (on the left, logarithmic scale) and percentage (right scale). From Pignatti 1984.

Table 5. Comparison of floristic and vegetational diversity among floras of marginal habitats, maquis and garrigue, and an evergreen oak forest

	marginal habitats	maquis and garrigue	evergreen oak forest
Diversity	low	high	low
Endemism	very high	high	low
Polyploidy	very low	high	low
Presence of annuals	absent	high	low

Theories about the origin of floras are in general highly speculative but in the Mediterranean vegetation useful information can be gained from the control of the degree of polyploidy. In the biological evolution the normal tendency is the passage from diploids to polyploids, the contrary was observed only in very few cases and always as the consequence of manipulations of the caryotype (*e.g.* in cultivated strains of *Beta*). A common indication is that species living in the secondary vegetation have high polyploidy; distinct species from the same genera live in marginal habitats: in this case they are mostly diploids. Also in climax vegetation diploids prevail, but in general the species occurring here are not related to those from the secondary vegetation.

Consequently we came to the hypothesis that the evolution of the Mediterranean flora can be considered as being strongly influenced by human action on the environment (Pignatti 1980, 1983, 1987). Under natural conditions in early Pleistocene the plant cover of the Mediterranean basin consisted of climax forests and dwarf shrub vegetation of marginal habitats; from the actualistic point of view both had a rather poor flora. Man started using fire ca. 200,000 years B.C. and burned vegetation for hunting; consequently a generalized impact on vegetation was possible. The evergreen climax forest - a highly conservative environment - was eliminated, new niches were created, and an intense process of speciation took place. Many of the species presently living in maquis and garrigue originated from the flora of marginal habitats through a reduction of their living cycle and the formation of polyploids. Vegetation types which we now consider as typical of Mediterranean landscapes, like maquis and garrigue, developed due to these processes.

It is important to point out that among the components of maquis and garrigue vegetation there are many edible plants, *e.g.* wild cereals (*Hordeum, Aegylops, Avena*), legumes (*Lupinus, Vicia, Lathyrus, Cicer*), bulbs (*Leopoldia, Allium*, orchids) and others (*Linum, Cynara*, other thistles, aromatic and medicinal plants). A positive feedback begins (Fig. 6).

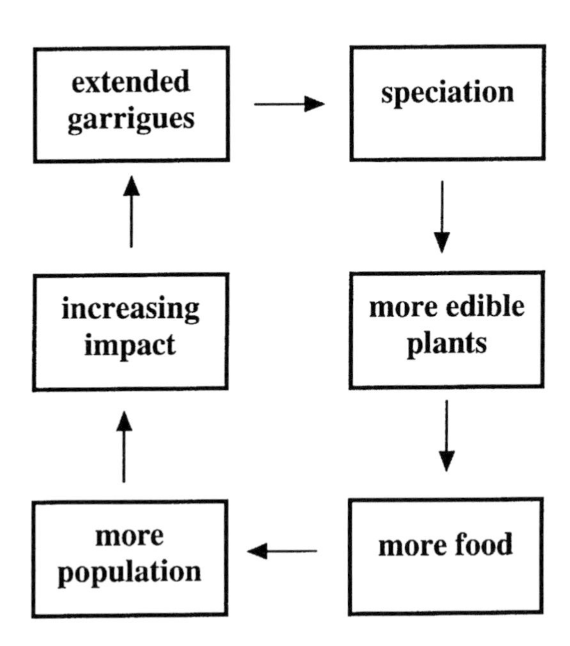

Figure 6. Positive feedback typical of Mediterranean landscapes.

Following tradition, men were hunters and women gatherers; but the hunting activity was very intensive, and after the last glaciation the stock of herbivores in Southern Europe became more and more reduced, so that in the palaeolithic culture only 10-15 % of food was supplied by meat (Cresta 1988) and the large majority by vegetables. It was the favourable condition for the passage to the activities of the Neolithic: hunters became shepherds whereas gatherers developed agriculture, the source of the Mediterranean civilization.

In this form, man stimulated the evolution of the flora, while plants in turn stimulated the cultural evolution of man.

Biodiversity conservation in the Mediterranean area

For the present time it is not only important to know how the Mediterranean flora and vegetation acquired a high degree of biodiversity but also to propose strategies for conservation. In the Mediterranean countries the landscape has been subject to heavy impact over large areas and in general reforestation is regarded as a very urgent task: is this the adequate strategy? Restoration of degraded areas under these climatic conditions can be obtained by use of pines (mainly *Pinus halepensis*) and of eucalypts (*Eucalyptus camaldulensis, E. globulus*), or by natural recovering of the woody vegetation. Many examples of reforestation have been investigated in Southern Italy and Sicily (Pignatti 1996). In about 20-year-old plantations, only common weeds (most of them annuals) occur under the cultivated trees and there is no sign of increasing biodiversity as to natural or semi-natural vegetation. In the Illyrian Karst near Trieste, under submediterranean conditions, large plantations of *Pinus nigra* with different ages of about 100-250 years are presently senescent and endangered by infestations and pollution; recovering of the natural vegetation was achieved only in few cases.

Natural recovering may give better results, and low-maquis vegetation can regenerate - at least in theory - in a time of 6-10 years and grow to a dense scrub, but in general such surfaces are exposed to repeated fires. Plantations seem preferable, because they are better protected from fire or other influences; in addition, the tree cover is important to stabilize the water balance and, after a period of some decades, will yield at least a small income. In any case, it seems that this remains in the general philosophy of monoculture, and may not represent the best way for restoration of a diversified plant cover.

Some examples of successful recovering may be discussed. In the Circeo National Park (south of Rome) on sandy soil, under the protection of a large plantation of *Pinus halepensis*, after 60 years a well-developed formation of evergreen Mediterranean shrubs is growing 2-3 m high; the trees have been cleared twice and will soon be totally removed; the vegetation will develop into the natural community of evergreen oaks (*Viburno-Quercetum ilicis*). In Sardinia near Siliqua a dense community of *Cistus monspeliensis* (acidocline shrubs spreading after repeated fires) was planted with eucalypts. Even if the wrong species were used, the vegetation has recovered very well; in about 10 years a high maquis has been stabilized and *Cistus* been mainly substituted by more resistant shrubs (*Phillyrea, Arbutus, Myrtus* and components of the final vegetation, as *Pistacia lentiscus* and *Quercus ilex*). Here reforestations with hard-wood eucalypts like *Eucalyptus marginata* would contribute to the creation of a new vital plant cover. Indeed, in both cases the final result is the restoration of the natural forest, which is good for landscaping and environmental protection, but diversity remains low.

As previously demonstrated, the high diversity of the Mediterranean vegetation is the result of environmental factors combined with the action of man. Restoring biodiversity implies a careful adaptation of sylvicultural measures to the ecological and social reality (Pignatti & Pignatti 1995). In any case, the use of trees will help, even if this does not represent a real solution. In the Basilicata region of Southern Italy there are large surfaces of Pliocene clay; here the horizontal surfaces (formerly covered by deciduous forests) are mostly cultivated with wheat, but a peculiar form of erosion locally indicated as "calanchi" hosts on small surfaces a very rich herbaceous vegetation (*Lygeo-Stipetalia*). Under such conditions recovering has to be planned with reduced areas of woodlands, surrounded by shrubby vegetation, and strict conservation of local fragments of the *Lygeum* association, only to prevent erosion and regulate the water cycle; this means, to re-create a diversified environment. A similar result was obtained by shepherds in Sardinia burning small patches of natural vegetation every year for millennia, in order to expand pastures: for the woody vegetation this is a heavy impact, but the diversity of flora and vegetation is preserved.

Conservation is a difficult task in Mediterranean countries and can be successful only when based on the preservation of traditional utilization forms of natural resources (Pignatti 1991). Consequently, a large-scale policy of reforestation, as foreseen by some EU directives, may combine high costs with negative results.

Acknowledgements

Research carried out with grants CNR and MURST 40 %. We are much indebted to Dr Laura Pignatti (Milano) for the careful revision of the English text.

References

Boissier, P.E. 1867-1884. Flora Orientalis. 5th vol., Basileae, Genevae, Lugduni.

Celesti, L. & Pignatti, S. 1988. Analysis of the chorological diversity in some South-European vegetational series. Annali di Bot. 46: 25-34.

Cresta, M. 1988. Ecologia umana. Ed. C.E.S.I., Roma.

De Lillis, M. & Testi, A. 1990. Post-fire dynamics in a disturbed Mediterranean community in Central Italy. pp. 53-62. In: Goldammer, J.G. & Jenkins, M.J. (eds), Fire in Ecosystem Dynamics. SPB Academic Publ., The Hague.

Feoli, E., Pignatti, E. & Pignatti, S. 1981. Successione indotta dal fuoco nel *Genisto-Callunetum* del Carso Triestino. Acta Biologica 58: 231-240.

Greuter, W., Burdet, H.M. & Long, G. 1984-89. Med-Checklist. Vol. 1, 3, 4. Conserv. Jard. Bot. Genève ed.

Haeupler, H. & Schönfelder, P. 1988. Atlas der Farn- und Blütenpflanzen der Bundesrepublik Deutschland. Ulmer, Stuttgart.

von Hayek, A. & Markgraf, F. 1924-1933. Prodromus Florae Peninsulae balcanicae. Feddes Rep. Beih. 30. 3rd vol.

Maire, R. *et al.* 1952-1980. Flore de l'Afrique du Nord. Vol. 1-15. Lechevalier, Paris.

Naveh, Z. & Whittaker, R. 1979. Structural and floristical diversity of shrublands and woodlands in northern Israel and other Mediterranean areas. Vegetatio 41: 171-190.
Perring, F.H. & Walters, S.M. 1962. Atlas of the British Flora. London.
Pignatti, G. 1996. Analisi vegetazionale e prospettive gestionali per impianti di arboricoltura da legno. PhD Thesis Univ. of Potenza, Italy.
Pignatti, G. & Pignatti, S. 1995. Land use and human influences in the evergreen broad-leaved forest regions of East Asia, the Mediterranean and Australia. pp. 199-210. In: Box, E.O. (ed), Vegetation Science in Forestry. Kluwer Acad. Publ., Amsterdam.
Pignatti, S. 1978. Evolutionary trends in Mediterranean flora and vegetation. Vegetatio 37: 175-185.
Pignatti, S. 1979. Plant geographical and morphological evidences in the evolution of the Mediterranean flora (with particular reference to the Italian representatives). Webbia 34: 243-255.
Pignatti, S. 1983. Human impact in the vegetation of the Mediterranean Basin. pp. 151-161. In: Holzner, W., Werger, M.J.A. & Ikusima, I. (eds), Man's Impact on Vegetation. Dr W. Junk Publ., The Hague.
Pignatti, S. 1984. The relationships between natural vegetation and social system in the Mediterranean basin. pp. 35-46. In: Miyawaki, A. (ed), Vegetation Ecology and Creation of New Environment.
Pignatti, S. 1986. The consequence of climate on the Mediterranean vegetation. Annali di Bot. 42: 123-130.
Pignatti, S. 1987. The relationships between natural vegetation and social system in the Mediterranean basin. pp. 35-45. In: Miyawaki A. *et al.* (eds), Vegetation Ecology and Creation of New Environments. Tokai Univ. Press, Tokyo.
Pignatti, S. 1991. Forest management and plant conservation in the Mediterranean area - Introduction. Botanika Chronika 10: 353-358.
Pignatti, S. 1995. Biodiversità della vegetazione mediterranea. Atti dei Convegni Lincei 115: 7-31.
Pignatti, S. in press. Diversity and uniformity of the Mediterranean flora. Ecologia Medit.
Poldini, L. 1991. Atlante corologico delle piante vascolari nel Friuli-Venezia Giulia. Amm. Reg. Udine.
Trabaud, L. & Lepart, J. 1980. Diversity and stability in garrigue ecosystems after fire. Vegetatio 43: 49-57.
Webb, D.A. 1978. Flora Europaea. Taxon 27: 3-14.

SPATIAL ARRANGEMENTS OF HABITATS AND BIODIVERSITY: AN APPROACH TO A SIGMASOCIOLOGICAL VIEW

ANGELIKA SCHWABE

Institut für Botanik der Technischen Universität / Geobotanik, Schnittspahnstraße 4, D-64287 Darmstadt, Germany

Key words: Central Alpine dry areas, diversity pattern, landscape sections, physiotopes, "steppe heath", vegetation complexes

Abstract

Results of biodiversity research are only applicable to the organization level under investigation. To analyze diversity at the level of the spatial arrangements of habitats it is necessary to investigate ecosystems/ecosystem complexes in landscape sections.

Since it is often impossible to determine ecosystems within their boundaries, we take physiotopes as plot areas (= topographical units in which homogeneous stable conditions in the substance regime prevail), which are geomorphologically characterized. In anthropogenically influenced landscapes, different types of land-use mark the boundaries of the plot areas, in addition to the physiotope factors. We apply the methods of sigmasociology taking vegetation complex relevés in the field.

Some case studies are presented showing regularities of diversity (*e.g.* calcareous rock physiotopes with "steppe heath", dry slopes of the Central Alpine dry areas, physiotopes of rivulet valleys).

The following regularities can be determined so far:

- The combination of micro-habitats with corresponding specific vegetation types follows a regular pattern.
- Anthropogenic influence often causes higher values of vegetation types in the investigated plots (this is in accordance with the "intermediate disturbance hypothesis"). If the influence is too strong the values drop again.
- Microgeomorphological heterogeneous physiotopes are colonized by a specific diversity pattern of plant communities, this pattern being quite often similar or nearly identical, even in large transects of several hundred km (example: "steppe heath" physiotopes of the Jura in Central Europe).
- Biocoenological investigations of the habitats of *Emberiza cia* (Rock Bunting) demonstrated *e.g.* that this bird species colonizes structure-analogously composed vegetation complexes in the Black Forest (Germany) and in the Grisons (Switzerland) with similar diversity patterns.
- In the Central Alpine dry areas the phytocoenosis diversity decreases in the case of reduced humidity and influence of continental climate.

A. Kratochwil (ed.), Biodiversity in ecosystems, 75-106

There are various application possibilities of the study of diversity by vegetation complexes, *e.g.* in the frame of monitoring and "Environmental Impact Assessments" (EIAs).

Introduction

If the importance of biodiversity with regard to the organizing hierarchies of biological systems is examined, as outlined *e.g.* by Odum (1983): starting from the molecule through to organelles, cells, organs, organ systems to individuals and from there on to populations, coenoses, and finally to ecosystems and landscapes, it becomes obvious that the term biodiversity can be applied to different hierarchical levels. Astonishingly though, this concept is not mentioned in the classical textbooks of botany (exception: Strasburger 1998), whereas nearly all ecologically oriented textbooks and dictionaries try to come to terms with it (*e.g.* Remmert 1984; Schaefer 1992; Tischler 1993; Wilmanns 1993). Already Stocker (1979a, b) included the concept of diversity ("Mannigfaltigkeit") in his 4-fold scheme of biology. He considered this to be a task to be tackled by taxonomists and ecologists.

After publication of the volume "Biodiversity", edited by Wilson (1988), which regards this topic mainly under the aspect of the diminishing diversity, *e.g.* through habitat destruction ("biodiversity crisis", see *e.g.* Stearns *et al.* 1990), a lot of articles dealt with the subject "biodiversity"; one symposium volume *e.g.* is about "Biodiversity and ecosystem function" (Schulze & Mooney 1993), another volume deals with the "Coexistence of species on changing landscapes" (Huston 1994). Solbrig (1991) offers scientific issues and research proposals, the "World Conservation Monitoring Centre" (1992) summarizes the "status of the earth's living resources", Hawksworth (1996) gives a synopsis of "measurement and estimation of biodiversity", and a very complex and differentiated view is presented by Heywood & Watson (1995). A critical comparison of some publications is given *e.g.* by Haeupler (in this volume). The different activities of the "United Nations Environment Program" are summarized in Heywood & Watson (l.c.).

The above-mentioned publications show that there are especially deficits in biodiversity research on the level of spatially heterogeneous ecosystems/ecosystem complexes or larger landscape sections ("systems diversity" in the sense of Heywood & Watson 1995: 26: "richness of ecological systems in a region or landscape").

There are deficits, too, in the comparison of defined ecosystem types over large transects of *e.g.* a few hundred kilometres. To define specific ecosystems they have to be typified, it is therefore necessary to analyze a lot of sample plots.

In this paper it will be shown that it is possible to typify the vegetational parts of ecosystems and their spatial arrangements with the help of sigmasociology. Sigmasociology (sigma = sum of vegetation types/plot area) describes an approach to analyze diversity for a number of sample plots with a size of approximately 0.1 to 2-3 ha/plot area (depending on the minimum area). Its aim is it to work out generalizations of the investigated ecosystems with their typified habitat composition.

After some remarks about hierarchical and organization levels of biodiversity as well as about methodological aspects of sigmasociology several case studies are presented. These studies show regularities of diversity in spatially and phenologically heterogeneous ecosystems and ecosystem complexes.

Of course, the study of the vegetational diversity is only a limited approach to the whole ecosystem, but vegetation *e.g.* as structural component, as resource for animals and microorganisms, for the monitoring of disturbances and successional changes (Huston 1994: 232) offers a valuable reference matrix. Numerous studies have confirmed the correlation of vegetation types (including their species composition) with the physical heterogeneity of the substrates and definable abiotic factors. Heywood & Watson (1995: 102) summarize in the chapter "Diversity in ecological systems": "We may also need to consider various aspects of the arrangement of these habitats and the heterogeneity of the landscape they produce. These spatial aspects are difficult to describe, but is increasingly recognized that landscape analysis of this type is important in understanding species diversity." Even the species concept is discussed *e.g.* concerning apomictic "species" which can grossly inflate measures of species biodiversity (Harper & Hawksworth 1996: 9).

The direction taken by Luder (1981) *e.g.* in the framework of the concept "landscape diversity", only lists units recognizable in aerial photographs (*e.g.* different area usage like fields, parks, river banks).

Hierarchical levels of biodiversity

As to the hierarchical levels of biological systems the diversity concept has to be differentiated and classified, according to the respective level. In various publications it is also associated with biochemical diversity (Feeny 1976; Odum 1983), the molecular genetic level (Templeton 1996), the multiplicity of physiological and biochemical adaptation strategies (*e.g.* Lüttge 1991), and genetic diversity (Stearns *et al.* 1990; Grabherr 1994; Frankel *et al.* 1995).

The following definitions of hierarchical levels are used according to Whittaker (1975, 1977), Heywood & Watson (1995) and Goetze & Schwabe (1997):

- alpha-diversity: Number of species of a definable size, in the phytocoenosis considered here the multiplicity of individual species forming a specific texture (Barkman 1979) and structure ("within-vegetation" after van der Maarel 1988; "species richness" after *e.g.* Grabherr 1994; "point-diversity" of one habitat: Mühlenberg 1993).
- beta-diversity: Gradients between different habitats *e.g.* along a given ecological gradient, proportion or rate ("between-community" after van der Maarel 1988, "gradient diversity" between habitats: Mühlenberg 1993).
- gamma-diversity: Diversity of larger regions, landscape-diversity, ecosystem types in a landscape, area diversity (s. *e.g.* Haber 1979; Luder 1981; Schaefer 1992). Often only regional species lists are given, in our case we investigate the diversity of vegetation types on the level of ecosystems and of ecosystem complexes.

  Zwölfer & Völkl (1993) who work with phytophagous organisms in inflorescences of Cardueae-species, interpret this term in a different way, *i.e.* as total number of phytophagous organisms in inflorescences. It is proposed by Goetze & Schwabe

(1997) to differentiate $gamma_1$-/$gamma_2$-diversity. For vegetation complexes this means: $gamma_1$-diversity: number of vegetation types within a vegetation complex, $gamma_2$-diversity: number of vegetation complexes within a landscape part.
- delta-diversity: In the phytocoenosis/vegetation complex: change in the number of vegetation types along an ecological gradient.

According to Whittaker (1977) and Heywood & Watson (1995) only in the case of alpha- and gamma-diversity these are "inventory diversities" with specific numbers *e.g.* of taxa or other "suitable measures", beta-diversity is a "differentiation diversity" measuring gradient.

Indices of diversity which show relations between species and individuals (*e.g.* Shannon-Wiener-Index) are not dealt with here. The value of such formulas often is low, only relative values can be assessed (Remmert 1984; Tischler 1993). High species numbers combined with varying individual numbers or lower species numbers combined with the same number of individuals result in equally high diversity values. Haeupler (1982, 1995, in this volume) has pointed out that the relativity of the Shannon-Wiener-Index has to be eliminated by evenness (equitability). But the following problem arises when working with evenness values: for a maximal possible diversity it is assumed that the species are equally distributed. Equal numbers of individuals for all species as an "orderly final state" are however only reached in specific cases, so there is often no real connection between evenness and definable vegetation types (see Haeupler 1982).

Deficits of results based on diversity indexes only are furthermore (examples):
- Important qualitative differences (*e.g.* occurrence of stenoecious species or such with a specific indicator value) which not necessarily become evident in higher abundance or coverage, are in most cases not considered.
- The importance of keystone species, which have essential effects on communities (see *e.g.* Lawton & Brown 1993; Bond 1993; Heywood & Watson 1995: 290 ff.; Harper & Hawksworth 1996), cannot be deduced from the indices.

The problem of the organization level and its investigation

Haeupler (1995) states that the more one goes into details and the more precisely one tries to compare, the less one becomes able to comprehend and to measure biodiversity. Only the taxa and their abundances which are part of the system structure are components that can easily be assessed (Haeupler l.c.). It will be shown in this contribution that in addition the syntaxa/vegetation types can be determined too.

As to the results of biodiversity, they are primarily valid only for the organization level being investigated. With increasing complexity of the system new system characteristics turn up, according to the principle of functional integration by Odum (1971). Directly connected therewith is the risk of reductionism: Regularities of a higher level of complexity cannot be detected even by the most intensive investigations on lower levels of complexity, since they have their own regularities as well.

The problem of biodiversity on a more complex higher hierarchical level has been defined by Körner (1993) as follows: "The fear of experimental failure at higher levels of complexity has led to a flood of papers of little relevance for the understanding of the behavior of diverse natural vegetation. Only in exceptional cases can functions at a subpopulation level be linked to ecosystem behavior."

In the preface to the symposium volume "Biodiversity and ecosystem function" it is stated that "this volume surveys the present state of knowledge about biodiversity" (Ehrlich 1993): a lot of basic data about diversity of ecosystems or landscape sections are however not yet available.

Methods used for the registration of diversity

Ecosystem and physiotope: the problem of plot demarcation in the field

Ecosystems and their boundaries have frequently been introduced schematically as being connected to influences like land use, history, climate etc.; quite often, publications include also biodiversity (see *e.g.* Schulze & Mooney 1993).

The main question concerning biodiversity at a higher level of complexity, *i.e.* the determination of ecosystems and their boundaries, can as yet hardly be answered. Ulrich (1993) considers the problem of determination rather pragmatically saying "The determination of ecosystems serves a purpose: it is the result of how the question has been formulated."

The boundaries of natural and seminatural ecosystems are *e.g.* in a relatively straight line if habitat factors are linear (*e.g.* flooding of lake- or river-banks); in other cases the limiting lines reveal as a rule a "fractality" of the boundaries.

In the case of forest landscapes it can, in general, be taken for granted - like in the classical areas for ecosystem research in the "Solling" mor beech forest (see *e.g.* Ellenberg *et al.* 1986) and in the "Göttinger Wald" mull beech forest (see *e.g.* Schaefer 1991) - that the forest communities also represent ecosystems: their determination does not seem to be too complicated. This is different in the case of "filigree" landscapes, where ecosystems consist of a mosaic pattern of microgeomorphologically different habitats. In order to be able to determine one has to know how to typify; the typifying being completed, modifications within similarities should be recognized. A multitude of different investigations is necessary to recognize singular as well as very characteristic types. This should be worked out in a time-saving way applying a field method.

As it is mostly impossible to determine the boundaries of ecosystems we take physiotopes as plot areas for vegetation complex relevés (see below). Leser (1996) and Thannheiser (1992) have used the term "geotope" as a synonym; it designates a unit with an approximate homogeneity which can be defined geomorphologically and topographically. According to Leser (l.c.) homogeneity is a relative quality and not essential for all parts of the system as a whole.

The following definition specifies the term "physiotope" after Neef (1981, modified): "in landscape science a topographical unit in which, as a result of the previous development, certain definable and often homogeneous stable conditions in the substance regime prevail (nutrient regime, water regime, air regime of the soil)".

In the case of complicated relief conditions, a physiotope can be divided into subtypes characterized by regularly arranged morphotopes (small geomorphological units). In the field and on maps, partly also on aerial photographs, physiotopes can, independent of their vegetation, be quite well differentiated, whereas it is only possible to correctly delimit biotopes when the relations of the biocoenoses living there are known. Thus geomorphologically differentiable physiotopes constitute the "components" (plot areas of vegetation complexes).

The term "ecotope" shall not be used in the following, as it is ambiguous. The definition given in a publication by Buchwald (1995) only takes into account the "spatial shaping of an ecosystem". Whittaker *et al.* (1973: 325) characterize an "ecotope" as follows: "The term is currently variously used as equivalent to habitat, biotope, microlandscape, or biogeocoenosis, but is not needed as a synonym of one of these. We suggest that henceforth it represents the species' relation to the full range of environmental and biotic variables affecting it." Also according to Schaefer (1992), the expression "ecotope" is used in very different ways.

In a certain physiotope type relief conditions of different small habitats are closely associated and form a pattern. Whether they represent regular combinations or not has to be explored. It can be postulated that a physiotype, *e.g.* a rock-complex with steep flanks, rock fissures, small shallows, and parts rich in fine earth has been built up by ever repeated combinations of habitats. An analysis of these combinations makes it possible to separate the irregular from the regular ones.

In anthropogenically influenced landscapes different types of land-use mark the boundaries of plot areas in addition to the physiotope factors.

The sigmasociological method

When taking relevés in a homogeneous area (dependent on the studied physiotope type the plot areas are approximately 1,000-20,000 m^2 in size), apart from geomorphological and other geographic details about the remaining abiotic and biotic environment, all vegetation units present (which must carefully be studied beforehand in order to make relevés), together with rankless units and single assessments of shrubs are listed in a 9-degree-scale (Table 1). With these relevés the occurring vegetation types represent a visible manifestation of the habitats and mini-habitats. The borders of such vegetation complexes can be determined by means of a minimum-area analysis similar to the classical phytosociological relevés (see *e.g.* Dierßen 1990; Dierschke 1994). Within a homogeneous vegetation complex, a state of saturation is attained when a certain number of communities has been reached; if the border to another complex type is crossed, the number and the specific quality of the vegetation types in the construction process change. In Fig. 1, two examples are presented: an extreme site complex of the inner Alps which is very hot and dry in summer and contains only an average of 13 vegetation units (minimum size of the area: 0.5 ha), as well as a more anthropogenically influenced type in the humid southern Black Forest (**Nardus stricta** grassland) comprising nearly 25 units (minimum area: 1 ha).

An overview can be gained by summarizing the relevés and drawing up a presence table (Table 2). The presence classes are identical with those of classical relevés (see *e.g.*

Dierschke 1994). Examples for presence tables of vegetation complexes are given, among others, by Dierßen (1979), Köppler & Schwabe (1996), Schwabe (1989, 1991a, b, c).

Table 1. Cover-abundance ratings of vegetation types in vegetation complex relevés according to Wilmanns & Tüxen (1978) and Schwabe & Mann (1990).

r	=	1	small stand
+	=	2-5	small stands or standard plots, cover < 5 %
1	=	6-50	small stands or standard plots, cover < 5 %
2m	=	> 50	small stands or standard plots, cover < 5 %
2a	=	any number, cover 5-12,5 % of the reference area	
2b	=	any number, cover 12,6-25 % of the reference area	
3	=	any number, cover ¼-½ of the reference area	
4	=	any number, cover ½-¾ of the reference area	
5	=	any number, cover more than ¾ of the reference area	

Upper limit for small stands:

Bryophyta/Lichenes communities	= 1 m^2
Therophyte/Hemicryptophyte communities	= 10 m^2
grassland/tall herbaceous/shrub communities	= 100 m^2
forest communities	= 1000 m^2.

For cover rates below 5% a big stand is separated into an adequate number of small stands ('standard plots').

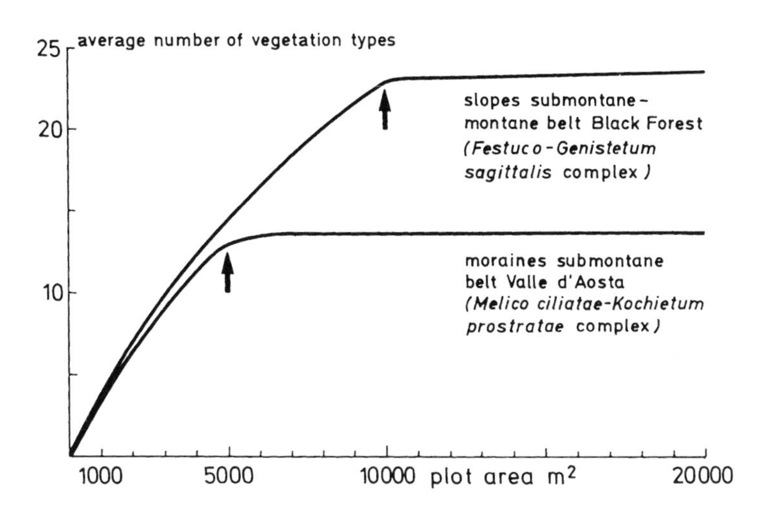

Figure 1. Minimum area (arrows) for 2 different vegetation complex types. The *Melico-Kochietum* is a near-natural *Festucetalia vallesiacae* association (Fig. 2, Table 3, column 1, see Braun-Blanquet 1961; Schwabe 1995), the *Festuco-Genistetum* complex (Black Forest, Vosges) shows an extensive anthropo-zoogenic influence ("intermediate disturbance hypothesis"; see chapter "Case studies" and Table 5, column 1).

Table 2. Presence classes of vegetation complex relevés

r	= present in less than	5 % of the compared plots
+	= present in	5- 10 % of the compared plots
I	= present in	11- 20 % of the compared plots
II	= present in	21- 40 % of the compared plots
III	= present in	41- 60 % of the compared plots
IV	= present in	61- 80 % of the compared plots
V	= present in	81-100 % of the compared plots

The denomination of the complexes is based on the characterizing communities. Preferentially for near-natural site complexes a unit of the potential natural vegetation is used to designate the complexes.

Fragments of different vegetation complexes are not always included in vegetation complex relevés. Vegetation complex investigations in urban or village regions, on anthropogenically influenced riverbanks, at railway stations etc. could only be performed considering fragment communities, which often serve as sensitive indicators. It is essential to standardize the inclusion of fragments.

The relevés obtained are classified according to similarity (for this purpose different, also multivariate procedures are available). By differentiating and characterizing vegetation units, the classified relevés can be assigned to types which often reveal landscape-typical distribution patterns.

The procedure is explained on the basis of examples taken from a dry region of the Central Alps: Fig. 2 represents a structural type, which can be observed in the moraine regions of the submontane-montane zone in the Valle d'Aosta: *e.g.* moraine tailings and habitats rich in fine earth, which in spring are to some degree influenced by needle ice. Some parts are consolidated, others, after rainfalls, are still subject to erosion. This micro-habitat differentiation is transparent through the vegetation mosaic; on the border, the needle ice areas are exclusively colonized by therophyte communities which reach their phenological optimum in March/April and are already dried out in May. The habitat pattern differs from the habitat complexes with rocks; the differences are evaluated with a number of vegetation complex relevés (Table 3).

The presented attempt of vegetation complex research has up to now been applied in Europe (*e.g.* Balcerkiewicz & Wojterska 1978; Dierßen 1979; Géhu 1991; Matuszkiewicz & Plit 1985; Pignatti 1978, 1981; Rivas-Martínez 1976; Schwabe, 1989, 1991b; Thannheiser 1988; Theurillat 1992a, b; Tüxen 1979; Vevle 1988), Canada (Ansseau & Grandtner 1990; Béguin *et al.* 1994), Africa (Deil 1995; Deil in press; Schwöppe & Thannheiser 1995), and Japan (Miyawaki 1978, 1980); for bibliographic review see Schwabe (1991a), for aspects of sigmachorology see Schwabe in press.

Another approach, including all synusia, is given *e.g.* by Gillet (1986), Gillet *et al.* (1991) and Gillet & Gallandat (1996). For special questions, *e.g.* concerning the diversity of such subsystems, the sigmasociological approach can be combined with the investigations of synusia. Basic research work for the connection between vegetation types and synusia was *e.g.* performed by Barkman (1973) and Schuhwerk (1986).

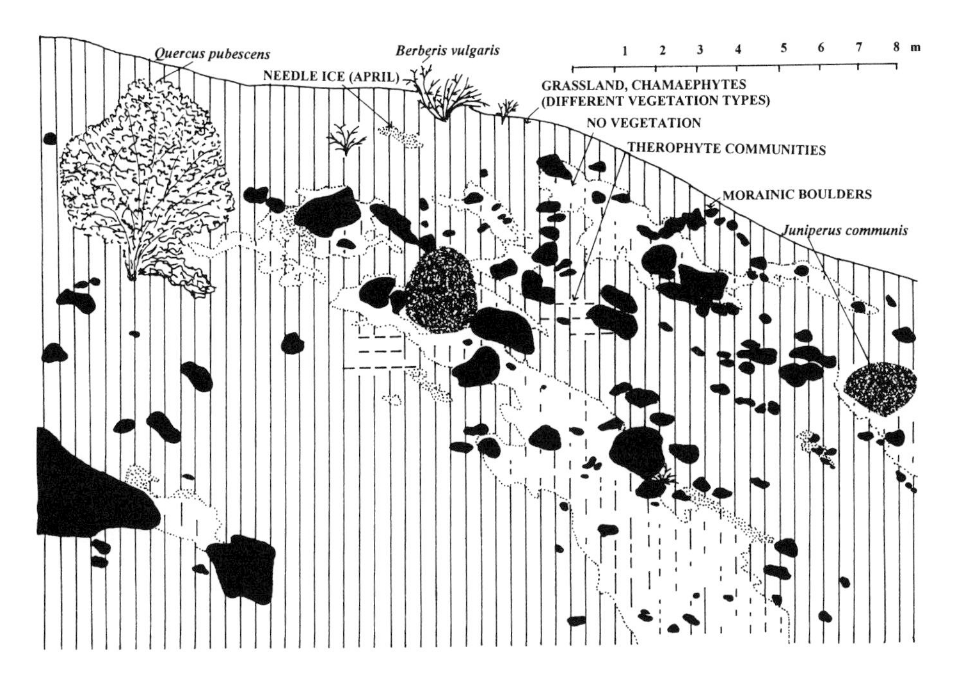

Figure 2. Moraine physiotope with different structural components (vernal aspect): consolidated parts, parts which are still subject to erosion with needle ice in winter and spring, shrubs (*Quercus pubescens*, *Juniperus communis*, *Berberis vulgaris* and others). Valle d'Aosta, Roisan: *Melico ciliatae-Kochietum prostratae* complex (11 different vegetation types, 800 m above sea level; see Table 3, column 1, Fig. 15).

Table 3. Part of a presence table of vegetation complexes: moraine and dry grassland/rocky physiotopes in the Valle d'Aosta (presence classes according to Table 2; exponents: amplitude of cover-abundance ratings according to Table 1)

physiotope number of vegetation complex relevés	moraine 8	rocks/grassland 12
cryptogamic synusia		
Fulgensietum fulgentis with *Psora decipiens*	$IV^{1\text{-}2m}$	$III^{+\text{-}2m}$
Polytrichum piliferum colony		$II^{+\text{-}1}$
Cladonietum convolutae		$+^{+}$
therophyte communities		
Alyssum alyssoides-Poa concinna community, rich in *Hornungia petraea*	$IV^{+\text{-}2m}$	$II^{1\text{-}2m}$
Alyssum al.-Poa concinna community	$V^{+\text{-}2m}$	$III^{1\text{-}2a}$
Arenaria leptoclados colonies	$V^{+\text{-}2m}$	$V^{+\text{-}2m}$
Trisetetum cavanillesii, only *Linaria simplex*	$II^{+\text{-}2m}$	I^{+}
Trisetetum cavanillesii	$+^{+}$	
Veronico-Poetum concinnae		$II^{2m\text{-}a}$
Veronico-Poetum con. with *Asterella saccata*		$II^{+\text{-}1}$

Table 3 (continued). Part of a presence table of vegetation complexes: moraine and dry grassland/rocky physiotopes in the Valle d'Aosta (presence classes according to Table 2; exponents: amplitude of cover-abundance ratings according to Table 1).

physiotope number of vegetation complex relevés	moraine 8	rocks/grassland 12
Veronico-Poetum con. with *Asterella saccata*		II^{+-1}
Trifolium arvense colony		II^{+-1}
Rumex acetosella colony		$+^{1}$
communities of rocky fissures		
Sedo dasyphylli-Asplenietum ceterach	I^{+}	III^{+-1}
Sedo das.-Aspl.cet., only *Sedum dasyphyllum*		II^{+}
Asplenietum trichomano-ruta-murariae		II^{+}
Asplenietum septentrionali-adianti-nigri		I^{+}
grassland communities		
Melico ciliatae -Kochietum prostatae	III^{1-3}	I^{+}
Melico cil.-Kochietum with *Agropyron intermedium*	IV^{+-3}	I^{+}
Onosmo cinerascentis-Koelerietum vallesianae	II^{3}	III^{+-3}
- *artemisietosum vallesiacae*	I^{2b}	I^{2a-b}
-- *Stipa capillata* variant	I^{+}	II^{2a-3}
- *phleetosum*		III^{1-3}
herbaceous margins		
Coronilla varia colony	I^{+}	$+^{+}$
Laserpitium siler hem communitiy		II^{+-2b}
Campanula bononiensis hem community		I^{+-1}
ground shrubs		
Arctostaphylos uva-ursi ground shrub	$I^{(+)}$	II^{+}
Juniperus sabina ground shrub		IV^{+-2a}
shrubs		
Prunus mahaleb shrub	IV^{+-1}	V^{+-2a}
Juniperus communis shrub	III^{+-2a}	IV^{+-2a}
Quercus pubescens shrub	IV^{+-2a}	II^{1-2a}
Celtis australis shrub	III^{+-1}	I^{1-2m}
Prunus spinosa shrub	$I^{(+)}$	II^{+-2a}
Rubus ulmifolius colony		II^{+-1}
Lonicera etrusca colony		II^{+-1}
Corylo-Populion		I^{+}

Case studies: diversity of selected physiotopes

Different sigmasociological case studies have shown that the combination of certain microhabitats in the vegetation complexes, on which again the diversity depends, does not occur at random but follows regularities.

Physiotopes of river and rivulet valleys

Using vegetation complexes, landscape-ecological classifications could be drawn up for rivers and rivulet valleys *e.g.* in the Black Forest (Schwabe 1987, 1989, 1991a, c). Fifteen main complex types were differentiated; Fig. 3 shows a section transect of the spatial arrangements of habitats of a rivulet bank of the Stellario-Alnetum glutinosae complex in the Black Forest. Similar classifications could be made in the Regnitz area (Bavaria, south-eastern Germany) by Asmus (1987) and in the Odenwald, western Germany, by Weißbecker (1993). In Fig. 4 a, b and c, a comparison of single vegetation complex relevés, presence classes of typified vegetation complexes and indicator values of the occurring vegetation types are presented.

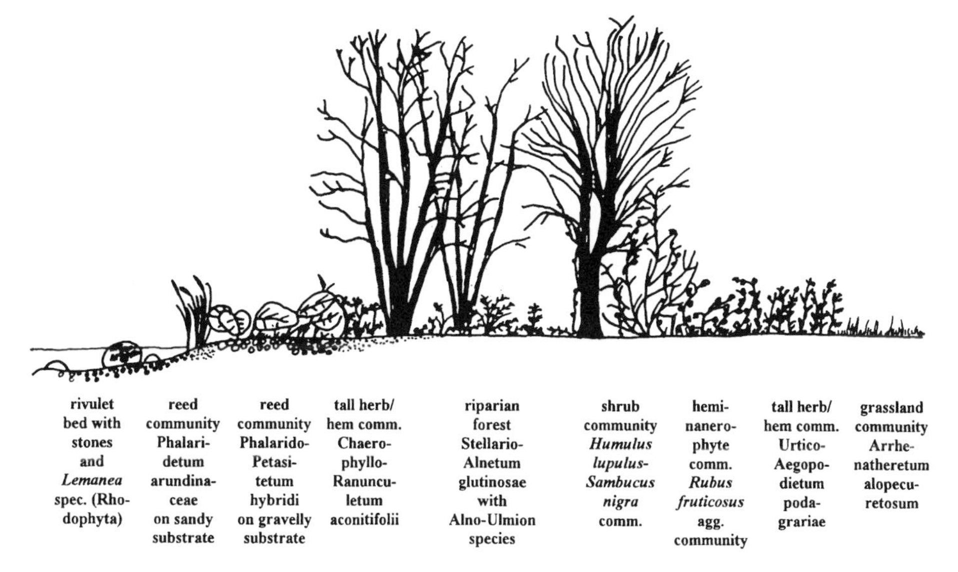

Figure 3. Spatial arrangements of habitats on a rivulet bank in the Black Forest (*Stellario-Alnetum glutinosae* complex); section of a transect.

Asmus (1987) and Schwabe (1987), performed - at the same time, but independent of each other - research at river bank vegetation complexes in regions of nearly identical sizes and found almost identical numbers of vegetation units present (Table 4).

If the plots to be compared have been subject to a strong anthropogenic impact, there is a decrease in vegetation units, in the case of extensive influences, an increase can be observed. The latter result is in accordance with the "intermediate disturbance hypothesis" (compare *e.g.* Rosenzweig 1995; Schaefer in this volume; Solbrig 1991). The near-natural vegetation complexes have lower numbers of vegetation units/plot area.

A direct consequence of these findings is that a high species- or community-diversity cannot be valued as positive in each case (compare *e.g.* Kinzelbach 1989: 68: "Diversity is no criterion of valuation for near-natural conditions"; Blab *et al.* 1995).

NEAR NATURAL (OLIGOHEMEROBIC) V-SHAPED VALLEYS

a)

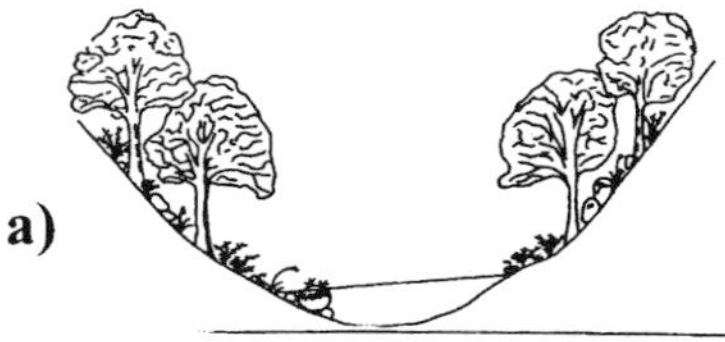

		one relevé	presence 10 relevés	different abiotic factors (examples), indicated by the specific vegetation types: climatic fact.						sedimentation					nutrients					velocity, dynamics				
				a	b	c	d	e	f	g	h	i	k	l	m	n	o	p	q	r	s	t	u	v
Forest comm.	Aceri-Fraxinetum	4	V^{1-5}	●											●								●	
	Stellario nemorum-Alnetum glutinosae	+	I^{+}				○								○									○
	Abieti-Fagetum	v	III	◑																				
Shrub comm.	*Rubus idaeus* community	+	V^{+-2}	●		●									●									
	Rubus fruticosus agg. community	.	II^{+-2}					⊙																
	Piceo-Sorbetum aucupariae	.	II^{+-2}																⊙					
Tall herb, fern comm.	Chaerophyllo-Ranunculetum aconitifolii	1	V^{+-2}	●												●								●
	Stachyo-Impatientetum noli-tangere	2	IV^{+-2}	●		●									●								●	
	Knautia dipsacifolia-Aegopodion comm.	+	IV^{+-1}	●											●									
	Athyrium filix-femina hem	2	IV^{+-1}	●		●																		
	Petasites albus colony	2	III^{1-2}	◑		◑										◑							◑	
	Urtica dioica colony	+	III^{+-2}												◑									
	Senecio ovatus colony	+	III^{+}																◑					
	Epilobium angustifolium colony	+	III^{+}																◑					
	Phalarido-Petasitetum hybridi	1	II^{1-2}		⊙					⊙					⊙									⊙
	Adenostyles alliariae-Athyrium filix-f. comm.	.	II^{+-2}	⊙		⊙																		
	Filipendula ulmaria comm.	.	II^{+}												⊙									⊙
	Luzula sylvatica comm.	.	II^{1-4}	⊙		⊙																		
Spring comm.	*Caltha palustris* colony	.	III^{+-1}								◑													
	Chrysosplenietum oppositifolii	.	II^{+}			⊙		⊙	⊙														⊙	
some Macro-phytes	*Lemanea* spec. (Rhodophyta)	v	II						⊙									⊙			⊙			
	Fontinalis antipyretica	v	+																					
				a	b	c	d	e	f	g	h	i	k	l	m	n	o	p	q	r	s	t	u	v

Figure 4. Single vegetation complex relevés, presence classes of typified vegetation complexes (shortened, complete table see Schwabe 1989) and indicator values of the occurring vegetation types. Presence classes: circle = +,1; circle with point = II, circle, partly black = III, circle, black = IV, V.a) *Aceri-Fraxinetum* complex, typical subcomplex, b) *Stellario-Alnetum* complex, submontane form, c) *Tanacetum-Convolvuletalia* community complex, typical subcomplex. Abiotic factors: -climatic factors: a) higher air moisture, b) partly sun irradiation, c) partly shaded, d) less than 150 frost days/year, e) habitats with mild winter conditions, f) low temperature of the flowing water -sedimentation: g) gravel transport, deposit, h) habitats rich in clay and/or detritus, i) new deposits of sediment, k) slight accumulation of sand, l) anthropogenically strongly influenced substrates (*e.g.* "Neopedon", murbruk) -nutrients, O_2: m) rich in nutrients (generally), n) habitat rich in bases, o) habitat rich in ammonium, p) habitats rich in O_2, q) intensive mineralization -velocity, dynamics: r) low velocity of the flowing water, s) high velocity of the flowing water, t) habitats with high groundwater level, u) habitats with much seepage water, v) periodical inundation.

b)

SLIGHTLY ANTHROPOGENIC INFLUENCED (MESOHEMEROBIC) FLOOD-PLAIN VALLEY

		one relevé	presence 11 relevés	different abiotic factors (examples), indicated by the specific vegetation types: climatic fact.						sedimentation					nutrients					velocity, dynamics				
				a	b	c	d	e	f	g	h	i	k	l	m	n	o	p	q	r	s	t	u	v
Forest comm.	Stellario nemorum-Alnetum glutinosae	3	V^{2-4}				●								●									●
Shrub comm.	Salicetum triandrae	1	I^{1-2}		○			○			○					○								○
	Rubus fruticosus agg. community	.	IV^{+-2}					●																
	Humulus lupulus-Sambucus nigra comm.	.	II^{+}					◑							◑									
Tall herb, fern comm.	*Impatiens glandulifera* colony	2	IV^{+-3}					●							●									●
	Urtica dioica-Convolvulus sepium comm.	3	IV^{2-3}					●							●									●
	Phalarido-Petasitetum hybridi	1	III^{+-2}		◑					◑					◑									◑
	Filipendula-ulmaria comm.	1	III^{+-2}												◑									◑
	Lysimachia vulgaris comm.	+	II^{+}					⊙																⊙
	Lythrum salicaria colony	+	II^{+}		⊙																	⊙		⊙
	Convolvulo-Epilobietum hirsuti	+	I^{+}		○			○							○									○
	Mentha longifolia colony	+	+											○										○
	Chaerophyllo-Ranunculetum aconitifolii	.	II^{+-1}	⊙		⊙*										⊙								⊙
	Knautia dipsacifolia-Aegopodion comm.	.	III^{+-1}	◑		◑*									◑									
Meadow comm.	Scirpetum sylvatici	+	I^{+}		○											○						○		
	Angelico-Cirsietum oleracei	+	+		○			○								○						○		
	Arrhenatheretum alopecuretosum	.	IV		●			●							●							●		
Pioneer comm.	*Agrostis*prorepens* comm.	+	III^{+}		◑							◑	◑							◑				◑
	Polygonum hydropiper comm.	+	II^{+}		⊙						⊙	⊙			⊙					⊙				⊙
	Veronica beccabunga colony	+	I^{+}		○						○									○				○
Reed swamp	Phalaridetum arundinaceae	2	IV^{+-3}		●																			●
	Iris pseudacorus colony	+	II^{+}		⊙			⊙							⊙					⊙				⊙
Macro-phytes/ comm.	Ranunculetum fluitantis	1	I^{1}		○										○			○						
	Callitrichetum hamulatae	.	II^{+-1}						⊙										⊙		⊙			
	Ranunculus trichophyllus colony/hybrid compl	.	II^{1-2}		⊙										⊙									
				a	b	c	d	e	f	g	h	i	k	l	m	n	o	p	q	r	s	t	u	v

STRONGLY ANTHROPOGENIC INFLUENCED (POLY-, EUHEMEROBIC) RIVULET VALLEYS

c)

		one relevé	presence 18 relevés	different abiotic factors (examples), indicated by the specific vegetation types																				
				climatic fact.						sedimentation					nutrients					velocity, dynamics				
				a	b	c	d	e	f	g	h	i	k	l	m	n	o	p	q	r	s	t	u	v
Shrub comm.	*Rubus fruticosus* agg. comm.	+	II^{+-1}				⊙																	
	Rubus caesius comm.	.	II^{+}				⊙							⊙		⊙								
Tall herb, fern comm.	*Impatiens glandulifera* colony	2	V^{+-3}					●							●									●
	Solidago gigantea colony	+	V^{+-1}		●			●						●										
	Tanacetum cuspidatum comm.	+	V^{+-3}		●			●						●										●
	Urtica dioica-Convolvulus sepium comm.	+	V^{+-2}		●			●						●	●									
	Mentha longifolia colony	1	IV^{+-3}					●							●									●
	Saponaria-Convolvuletalia comm.	+	IV^{+-1}		●									●										●
	Lythrum salicaria colony	+	IV^{+-1}		●			●						●		●								●
		+	IV^{+}		●																	●		●
	Aster lanceolatus-Senecion fluv. comm.	2	III^{+-2}		◐			◐						◐										◐
	Lysimachia vulgaris comm.	+	III^{+}					◐																◐
	Mimulus guttatus colony	+	III^{+}		◐																			◐
	Filipendula ulmaria comm.	+	II^{+-2}												⊙									⊙
	Convolvulo-Epilobietum hirsuti	+	II^{+-2}		⊙			⊙							⊙									⊙
	Urtica dioica colony	.	II^{+-2}												⊙									
	Phalarido-Petasitetum hybridi	.	II^{+-2}		⊙					⊙					⊙									⊙
Meadow comm.	Arrhenatherion fragment comm.	v	IV		●																			
Pioneer comm.	*Agrostis*prorepens* comm.	+	IV^{+}		●							●	●							●				●
	Equisetum arvense comm.	+	III^{+}		◐							◐												
	Polygonum hydropiper comm.	+	II^{+}		⊙						⊙	⊙			⊙					⊙				⊙
	Agropyro-Rorippetum austriacae	+	I^{+-1}		○			○			○	○								○				○
reed swamp	Phalaridetum arundinaceae	3	V$^{1-4}$		●																			●
	Phragmites australis colony	+	+$^{+}$		○										○					○				○
Macro-phytes	*Ranunculus peltatus* colony/hybrid compl.	.	II^{+-1}																	⊙				
				a	b	c	d	e	f	g	h	i	k	l	m	n	o	p	q	r	s	t	u	v

Table 4. Comparison of biodiversity (in this case number of vegetation types) on brook/river banks:
1. Region of river Regnitz (Bavaria) sensu Asmus (1987)
2. Black Forest sensu Schwabe (1987)

Asmus 1987	Schwabe 1987
river region of Regnitz northern Bavaria area size: 7545 km² 81 different vegetation types belonging to 15 phytosociological classes	Black Forest area size: 6000 km² 80 different vegetation types belonging to 16 phytosociological classes

A new approach aims at typifying entire landscape sections and their habitat arrangements (example: rivulet valleys in south-western Germany) and at determining gamma-biodiversity in dependence on physico-geographical and anthropo-geographical factors. The investigated cross sections of the valleys consist - as a rule - of 8 different vegetation complexes with often 50 different vegetation types (Goetze 1996; Goetze & Schwabe 1997).

Rock physiotopes

Habitats of Emberiza cia. Schwabe & Mann (1990) investigated grassland/rock-complexes, colonized by the Rock Bunting (*Emberiza cia*). The average size of the plot areas amounts to about 3 ha. Relevés from the Black Forest and the southern Alps show the different structural types of vegetation occurring in *Emberiza cia* habitats (Table 5). Structurally analogous plant communities replace one another. A simplified triangular depiction reflects the relations (Fig. 5).

The nesting sites are characterized by a highly differentiated and spatially heterogeneous structure; part of an *Emberiza cia* habitat is shown in Fig. 6 ("patchiness in space"). The "patchiness in space" can be typified with the vegetation complex relevés, afterwards it is possible to differentiate pattern types.

The "patchiness in time" has been studied considering the structural profile in different seasons and the phenological development of the plot areas. As there are close connections between the phenology of the vegetation and the food resources of *Emberiza cia* (fruit phase, animals, *e.g.* Orthoptera and caterpillars of different species especially for the nestlings), the phenological phases are important (Schwabe & Mann 1990).

Transect from the French and the Swiss Jura to the Swabian and Franconian Jura. In a transect from the French and the Swiss Jura to the Swabian and Franconian Jura, southern- and western-exposed calcareous rock areas were analyzed by vegetation complex relevés. Altogether 72 complex relevés, each covering an area of 0.1 to 0.4 ha, could be sampled (see as example for the physiographic character of the plot areas Fig. 7, 8).

Table 5. Presence table of *Emberiza cia* habitat types in the Black Forest
(two different vegetation complexes with 4 structural types of vegetation, partly replacing each other)
<< = indicator communities for mild winter conditions (subatlantic distribution)
<<<< = indicator communities for warm (partly hot) summer conditions

Structural type		**Festuco-Genistetum sagittalis-compl.**	**Calamagrostis arundinacea-Senecio ovatus-comm.-compl.**
	Number of vegetation complex relevés	18	5
	Average number of vegetation types	28,4	23,6
STRUCTURAL TYPE 1	**Grassland comm. without fertilizer influence:**		
	Festuco-Genistetum sagittalis typicum:		
	-- *Hieracium pilosella* facies	V +-2	+ +
	-- *Deschampsia flexuosa* facies, fallow land	IV 1-3	+ +
	-- *Calluna vulgaris* facies	V +-2	.
	-- *Holcus mollis* facies	II +-2	.
	-- *Deschampsia flexuosa* facies, extensively grazed	I 2-3	.
	-- *Dianthus deltoides* facies on gritty soil	I +-r	.
	Grassland comm. with low influence of fertilizer:		
	Festuco-Genistetum sagittalis trifolietosum		
	-- Typical form, extensively grazed	III 1-3	.
	-- Typical form, fallow land	II 1-3	.
	-- *Hypericum perforatum* facies	II +-2	.
	-- *Holcus mollis* facies	I +-1	.
	Agrostis capillaris-Festuca nigrescens grassland	II 1-3	.
	Poa chaixii colonies	I 2m	.
	Geranio-Trisetetum, fallow land	I 1-2	.
	Grassland of clearings:		
	Deschampsia cespitosa colonies	+ +	III +-2
	Carex pilulifera colonies	+ +	I 2
	Calamagrostis arundinacea-Senecio ovatus comm., only fragments	II +-1	.
	Calamagrostis arundinacea-Senecio ovatus comm.	.	V 1-4
	Luzula sylvatica colonies	.	III +-1
	Agrostis capillaris clearing grassland	.	III +-1
	Deschampsia flexuosa clearing grassland	.	II 2m
STRUCTURAL TYPE 2	**Pioneer vegetation, rock fissure vegetation:**		
	Sileno rupestris-Sedetum annui (fragment ''')	V +-2	+ +'''
	Rumex acetosella pioneer vegetation	V +-1	I +
	<<Galeopsietum segetum	IV +-2m	I 1
	Linaria vulgaris colony	III +-1	I 1
	Gymnocarpium dryopteris colony	I +	I 2m
	Senecio viscosus colony on rock waste	+ r	I +
	Rhacomitrium canescens synusium	V +-2m	.
	Woodsio-Asplenietum septentrionalis	III +-1	.
	ant hills with *Thymus pulegioides*	III +-1	.
	<<*Jasione montana*-pioneer vegetation	III +	.
	Vaccinium myrtillus colony	II r-1	.
	rocks with *Lasallia pustulata*	II +-2m	.
	<<<< *Rhytidium rugosum* synusium	II +-1	.
	Cladonia arbuscula colony	II +-2m	.
	Sedum telephium s.str. colony	I +	.
	Veronica fruticans colony	I +	.
	<<<< *Festuca ovina* agg. vegetation on rocks	I +-1	.
	ant hills with *Thymus froelichianus*	I 2m	.
	<< *Scleranthus perennis-Ornithopus perpusillus* comm.	I r-+	.
	Baemomyces roseus synusium	I +	.
	Polytrichum piliferum colony	I +-1	.
	Asplenium trichomanes colony	I r-+	.
	Thelypteris phegopteris colony	I +	.
	Veronica officinalis vegetation	.	III +-1
	Galeopsis tetrahit colony on gritty soil	.	III +-2m
	Polytrichum formosum cushion on rock waste	.	III +-1
	layers of *Hypnum cupressiforme*	.	II 2m

		Festuco-Genistetum sagittalis-compl.	Calamagrostis arundinacea-Senecio ovatus-comm.-compl.
	Number of vegetation complex relevés	18	5
	Average number of vegetation types	28,4	23,6
STRUCTURAL TYPE 3	**Ferns/ Chamaephytes:**		
	Vaccinium myrtillus vegetation	III +-2m	III +-1
	Athyrium filix-femina hem	III +	I +
	Pteridium aquilinum dominance stand	III 1-3	.
	Tall herb hem communities:		
	<< *Teucrium scorodonia* hem	V +-2m	V +-2m
	Urtica dioica hem	II r-+	II +-1
	<<<< *Trifolium medium* hem	II +-1	I +
	<< Teucrio-Centaureetum nemoralis	II +-2m	.
	Calamintha clinopodium hem	II +-1	.
	<<<< *Origanum vulgare* hem	I 1	.
	Agrimonia eupatorium hem	I +	.
	<<<< *Vincetoxicum hirundinaria* hem	+ +	.
	Communities of forest clearings (herbs dominant):		
	Fragaria vesca vegetation	V +-2m	IV 1
	Hypericum perforatum colony	IV +-1	III +-1
	Senecionetum fuchsii (= ovati)	I +-2m	IV 2
	Epilobium angustifolium colony	I r-+	V +-2m
	Galium harcynicum colony	I +-1	III +-1
	<< *Digitalis purpurea* colony	I +-1	III 2m
	Gnaphalium sylvaticum colony	I +	III +-1
	Galeopsis tetrahit vegetation	I +	II +
	Calamagrostio-Digitalietum grandiflorae	I +-1	I +
	Senecio ovatus colony	III +-1	.
	Senecio hercynicus colony	.	I 2m
STRUCTURAL TYPE 4	**Hemi-Nanophanerophytes dominant:**		
	Rubus idaeus vegetation (on gritty soil)	V +-2	IV 2
	Rubus fruticosus agg. vegetation	IV +-1	IV 1
	<<<< *Rubus canescens* comm. (upwards 1100 m NN!)	III +-2	.
	<<<< *Rubus bifrons* comm.	II +-2m	.
	<< *Rubus pedemontanus* veg. (upwards 1260 m NN!)	.	I 1
	Nanophanerophytes dominant:		
	Fagus sylvatica shrub	II 1-2	II 1
	Sambucetum racemosae	I +	II +-2
	Pseudotsuga menziesii (young afforestation)	I 2	II 2
	Picea abies (young afforestation)	+ 1	II 2
	Sarothamnus scoparius shrub	+ 1	I 1
	Rosa pendulina comm.	+ +	I +
	<<<< Pruno-Ligustretum	IV +-2	.
	Fagus-Acer pseudoplatanus-Corylus shrub	II r-2	.
	Picea abies (groups of young trees)	I 1-2	.
	Corylus avellana shrub	I +-2	.
	<<<< Cotoneastro-Amelanchieretum (species poor)	I r	.
	Crataegus monogyna shrub	I 1	.
	Salix caprea shrub	.	I +
	Forest communities:		
	Luzulo-Fagetum and *Fagus*-stand	II 1-2	.
	Alnus glutinosa stand	I +-2	.
	old *Picea abies* afforestation	+ 1	.
	Other communities:		
	Lupinus polyphyllus colony (sown)	+ 1	III +
	Rumex obtusifolius colony	+ +	I +
	Epilobio-Juncetum effusi	II +-1	.
	Parnassio-Caricetum fuscae	I +	.

<<<< = indicator community: warmth in summer
<< = indicator community: subatlantic conditions (mild winter)

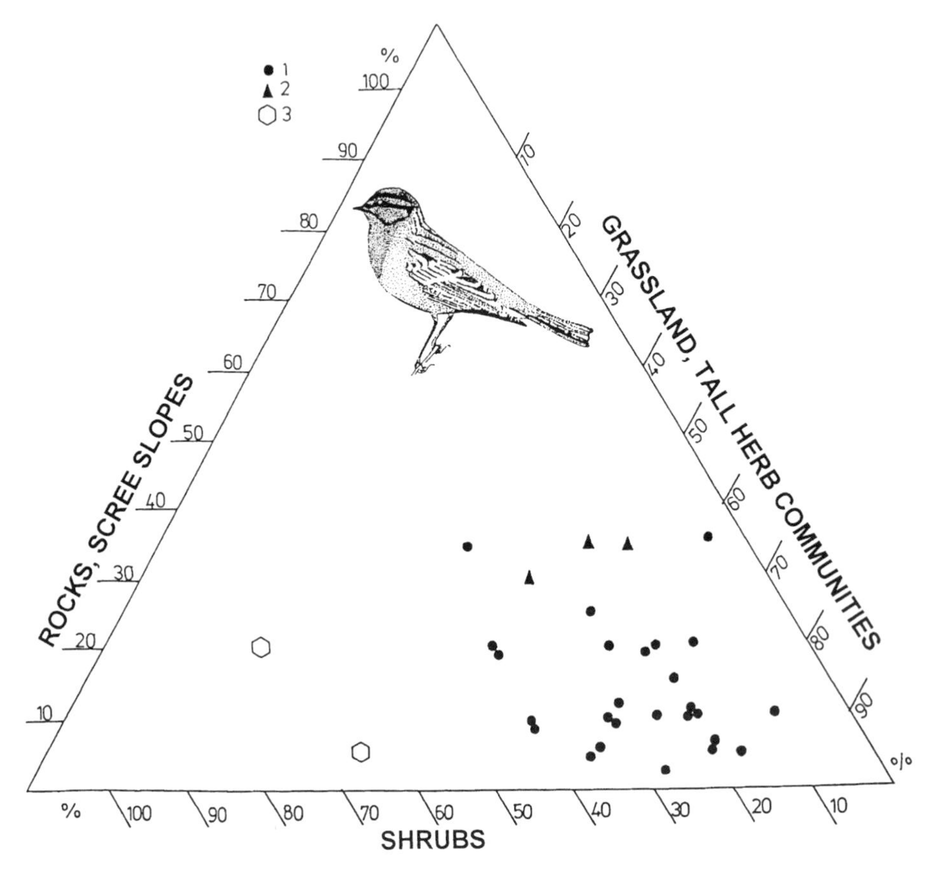

Figure 5. Triangular scheme of the proportion: grassland, tall herb community, rocks and shrubs in *Emberiza cia* habitats. 1 = Black Forest; 2 = the Grisons, 3 = former breeding sites of *E. cia* (Black Forest); see also Table 5.

The differentiated vegetation mosaic of the Jura "steppe heaths" in a broader sense is classified and typified by the complex relevés (original relevés: see Köppler 1995). Complex groups, defined by a structural similarity of vegetation complex types, could be classified (Fig. 9, 10).

In all investigated landscape areas, 21 complex types could be distinguished (Köppler 1995; Köppler & Schwabe 1995; Schwabe *et al.*, 1992).

The vegetation types drawn up for each relevé area can be assigned to the formations: fissure vegetation, pioneer vegetation, groups of woody species in rocky areas, grassland communities, fringe communities, shrubs. Thus it is possible to define formation patterns (diversity patterns) of the complex types. In a concentrated form, the richness of the vegetation mosaic is shown, indicating the average values of the vegetation types per complex type (Fig. 11).

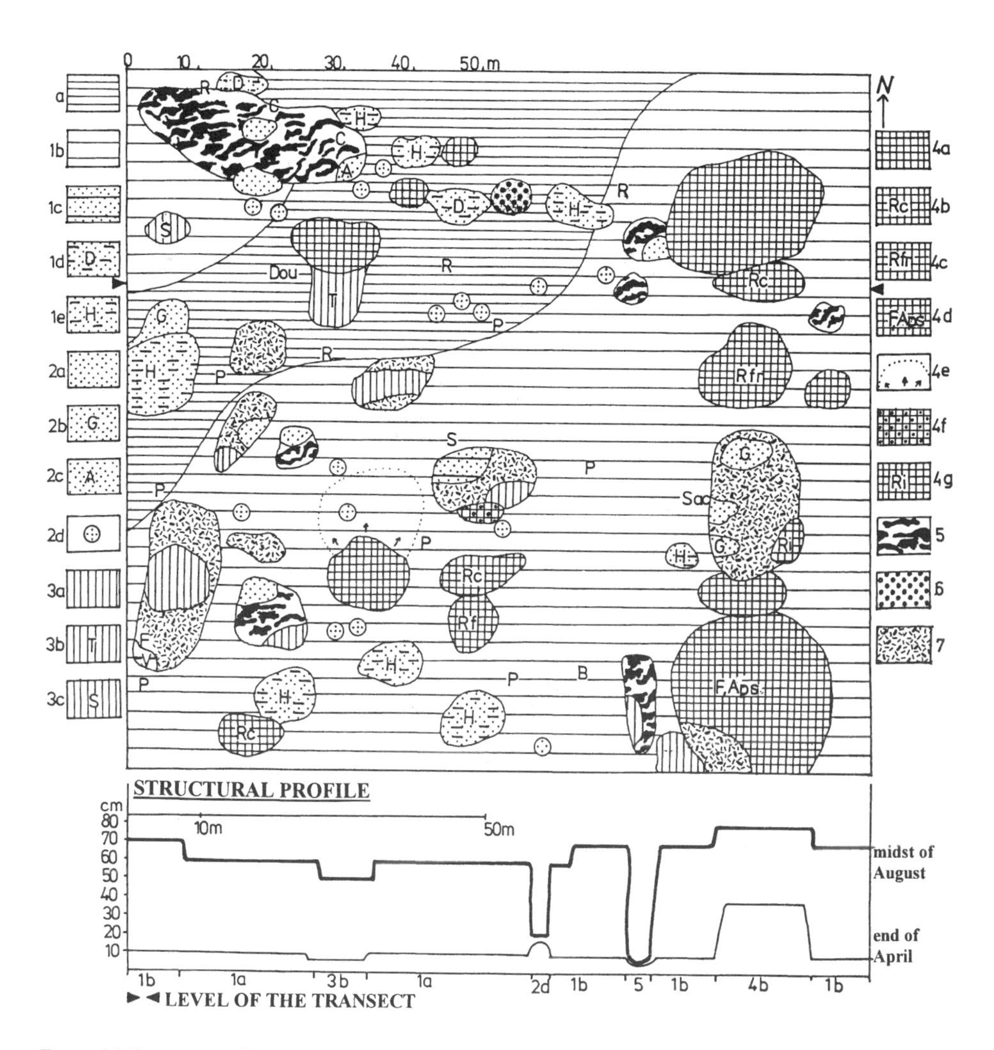

Figure 6. Micropattern of an *Emberiza cia* breeding site (Black Forest, Tunau, 850 m m.s.l.), southern-inclined, showing the spatial arrangements of the microhabitats of the vegetation. It is typified in Table 5, column 1. 1 Grassland communities: *Festuco-Genistetum sagittalis*, 1a - *trifolietosum*, typ. form, fallow land, 1b - *typicum*, *Deschampsia flexuosa* facies, fallow land, 1c - *typicum*, *Holcus mollis* facies, 1d - *typicum*, *Dianthus deltoides* facies, 1e — *typicum*, *Hieracium pilosella* facies, 2 Pioneer vegetation, vegetation of rocks, 2a *Sileno-Sedetum annui* and general signature for 2, 2b *Galeopsietum segetum*, 2c *Woodsio-Asplenietum septentrionalis*, 2d ant hills with pioneer vegetation, 3 Tall herb hem communities, vegetation of forest clearings, 3a *Teucrium scorodonia* hem, 3b *Trifolium medium* hem, 3c *Senecio fuchsii* colony, 4 Shrubs and hemi-nanophanerophytes, 4a *Pruno-Ligustretum* (typical and fragments), 4b *Rubus canescens* comm., 4c *Rubus fruticosus* agg. vegetation, 4d *Fagus-Acer pseudoplatanus-Corylus* shrub, 4e areas with young *Prunus spinosa* clones, 4f *Prunus spinosa*, small scrubs, 4g *Rubus idaeus* vegetation, 5 rocks, 6 fragments of bedrock, 7 scree areas. Single shrubs: Sac *Sorbus aucuparia*, S *Sorbus aria*, P *Picea abies*, B *Betula pendula*, C *Crataegus monogyna*, R *Rosa canina*, F *Fraxinus excelsior*, Dou *Pseudotsuga menziesii*, V *Vaccinium myrtillus*.

Figure 7. Part of the investigation area in the upper Danube valley near Beuron: the rock physiotopes indicate the plot areas of the vegetation complex relevés. 600-700 m m.s.l. Photo: D. Köppler.

Figure 8. Part of a plot area in the upper Danube valley: *Pulsatillo-Caricetum humilis* complex with *e.g. Geranio-Peucedanetum cervariae*, *Cotoneastro-Amelanchieretum* (Type 3a1.1, see Figure 10). "Rauher Stein" near Beuron (upper Danube valley). 770 m m.s.l. Photo: D. Köppler.

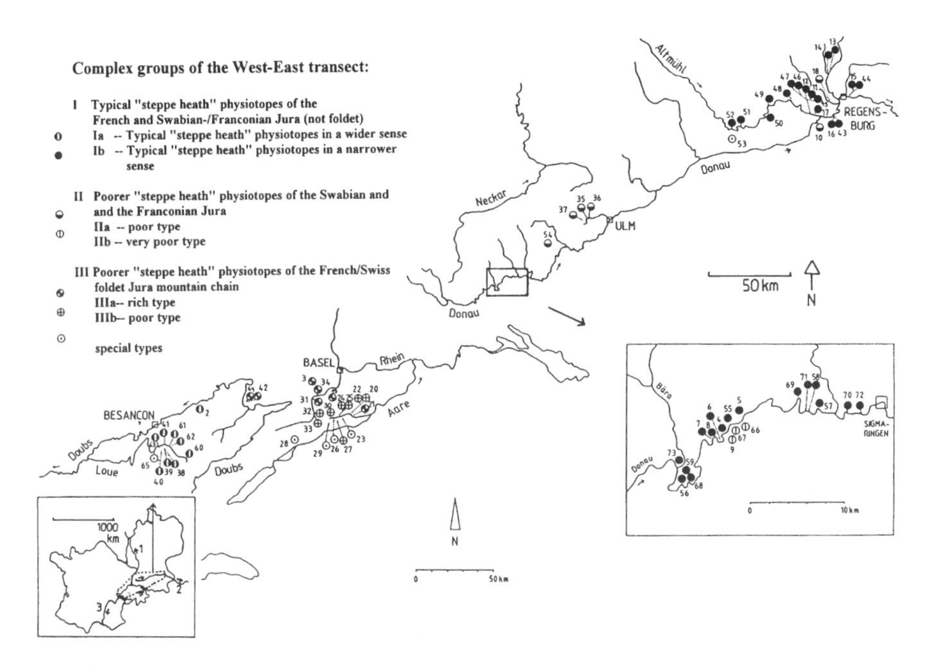

Figure 9. Plot areas and differentiated complex groups of the "steppe heath" vegetation complexes in the French, Swiss, Swabian and Franconian Jura (compare Köppler 1995, Köppler & Schwabe 1996). Small map: Boundaries of France, Switzerland and Germany with the investigation area: 1 = Rhine, 2 = Danube, 3 = Rhône.

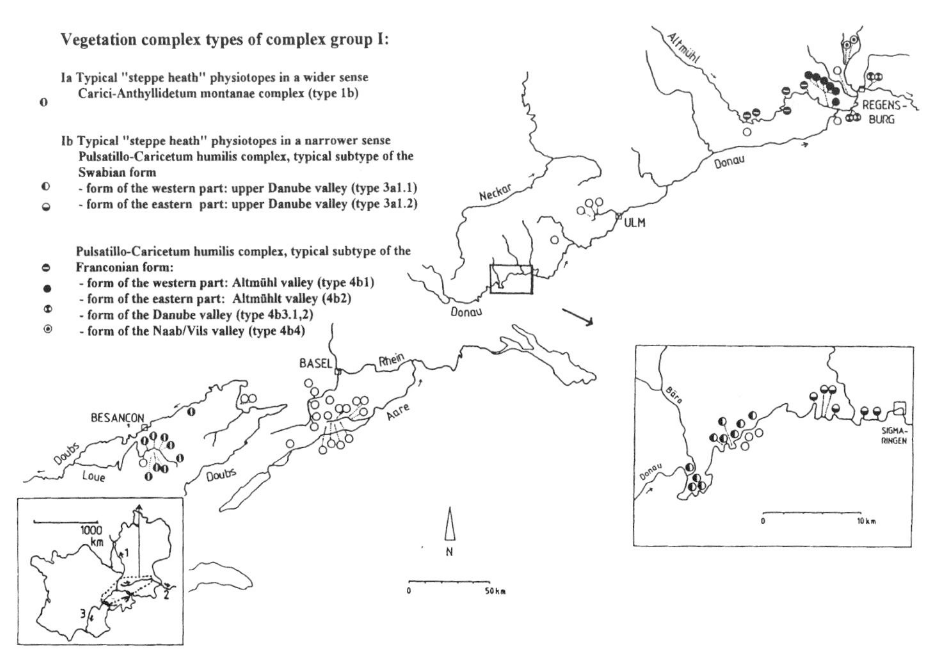

Figure 10. Differentiation of the "typical steppe heath complexes" (compare Köppler 1995; Köppler & Schwabe 1996), abbr. see Figure 9.

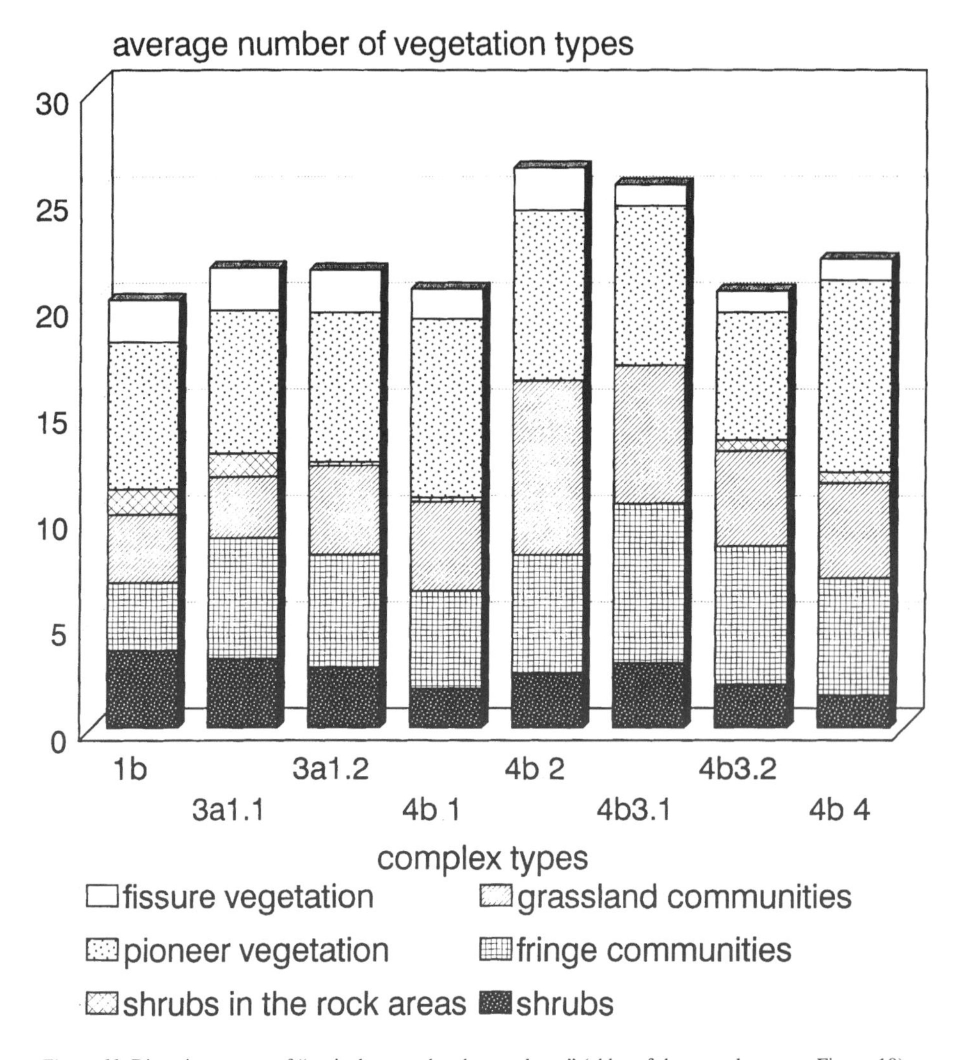

Figure 11. Diversity pattern of "typical steppe heath complexes" (abbr. of the complexes see Figure 10).

There are distinct gradients of these patterns from west to east, *e.g.* in the substitution of shrub vegetation types - which play an important rôle in the submediterraneanly influenced areas - by fringe communities (*Trifolio-Geranietea*) in the subcontinentally influenced regions in the east (Fig. 11).

According to the results of the vegetation complex investigations, we can define Jura "steppe heath" physiotypes on the level of their formation pattern and their diversity of vegetation types as follows: "Typical Jura 'steppe heath' physiotopes in a broader sense" are rock site complexes in Central Europe, which are rich in structures and show a typifiable mosaic of xerophytic vegetation consisting structurally of vegetation units which belong to the following 5 formations:

1. fissure vegetation
2. pioneer vegetation
3. grassland communities
4. fringe communities
5. shrubs.

When the vegetation types are subdivided down to facies level, the mean numbers of the vegetation types of one complex type amount to about 20 or more (relative value). The vegetation complex types with their corresponding diversity patterns can be assigned to several complex groups. These groups are characterized by a comparatively great uniformity of structure and by quite similar average numbers of existing vegetation types.

Rock- and moraine-physiotopes in central alpine dry areas. The comparison of large areas shows also for dry habitats in the Central Alps (Valle d'Aosta, Valais, Vinschgau, Valtellina, Engadine, Upper Inn valley, Rhine valley of Chur) that the "licenses" for microhabitats of certain physiotope-types in large-area transects correspond to one another. The vegetation types are - if not identical - analogous in structure. From the large-area comparison a conformity of the diversity patterns was worked out, the result of which proves the biocoenotic basic rules of Thienemann (1920, 1956) also on the level of complex ecosystems (Schwabe & Kratochwil 1994). In dependence on the xerothermy of the habitat complexes and the decrease in the yearly precipitation, the number of phytocoenoses per investigation area between the mouth of the valley and its centre increases once more with raising sea level and humidity in the upper montane zones (Fig. 12, 13, 14, 15, 16). This regularity can be superimposed by intensive anthropo-zoogenic influences, leading to a decrease in communities (example Vinschgau). In all investigated valleys higher humidity or the impact of warm wind (foehn) are a precondition for the occurrence of a multiplicity of phytocoenoses. Higher humidity in the investigation area enhances above all the occurrence of fringe communities of *Trifolio-Geranietea sanguinei* (Fig. 12, 13), which are missing in most xerothermic vegetation complexes on sunny spots. Thus we can define the length-profile of an "ideal" central alpine valley for the southern slopes by means of the phytocoenosis diversity. For areas with altogether lacking or extensive anthropogenic impact, the curve has the same form as the one showing the distribution of the precipitation. In the upper montane-subalpine zone the curve declines again regularly and is no longer congruous with the distribution of the precipitation. Frost and the short vegetation period have such an extreme effect that the inventory of therophyte-floras, shrub communities and others *e.g.* is only minor or lacking altogether.

This can be formulated analogously with regard to the second biocoenotic basic principle as coenologic-landscape-ecological principle (Schwabe & Kratochwil 1994): "The more the living conditions of a habitat complex deviate from the normal - and, as regards most coenoses/synusia, from the optimum - the greater being the loss of coenoses/synusia; however, the more characteristic and abundant will they become."

The hypothesis of Elton (1933) "the community really is an organized community in that it has 'limited membership'" is transferable to the level of vegetation complexes too: the investigations *e.g.* in the inner Alps and in the "steppe heath" regions have shown that there is a "limited membership" for definable plant communities in definable vegetation types.

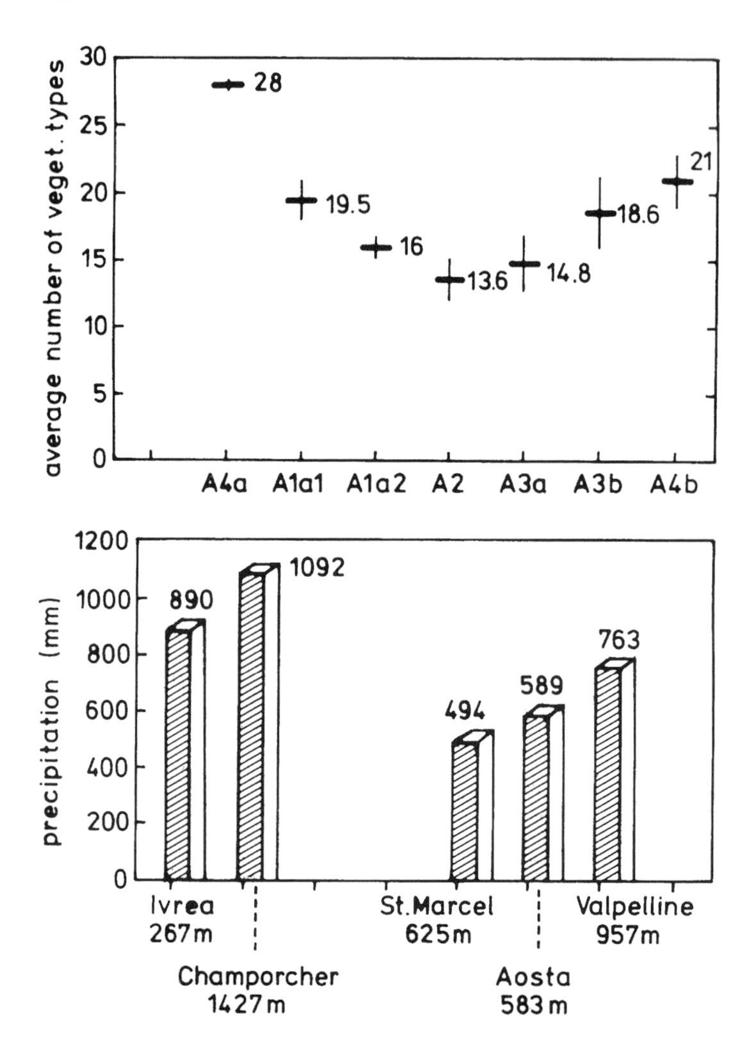

Figure 12. Average number of vegetation types/plot area for different complex types of the Valle d'Aosta (resp.plot area nearly 5,000 m² or 0.5 hectares; current of the river Dora Baltea from right to left) and annual precipitation (mm/a) after Walter & Lieth (1960 ff.). Vegetation types (for Type A2, A3 see Table 3)

A1a1: *Thymus vulgaris-Bromus erectus* community (*Alyssum argenteum*-form) complex
A1a2: *Thymus vulgaris-Bromus erectus* community (typical form) complex
A2: *Melico ciliatae-Kochietum prostratae* complex
A3: *Onosmo cinerascentis-Koelerietum vallesianae* complex
A3a: submontane-montane form
A3b: montane form s.str.
A4: *Pulsatillo montanae-Brometum erecti* (*Trinia glauca*-vicariant) complex
A4a: form of the more humid regions
A4b: typical form

The precipitation values are ordered according to the topographical sites of the vegetation types.

Figure 13. Type A4a in the Valle d'Aosta (example): *Pulsatillo montanae-Brometum erecti* (*Trinia glauca*-vicariant) complex, form of the more humid regions. The complex is *e.g.* characterized by tall herb hem communities, *Corylus avellana* shrubs and higher diversity of vegetation types. 1,300 m m.s.l., SSW, near Pta. Zuccore. Photo: A. Schwabe.

Figure 14. Type A1a2 in the Valle d'Aosta (example): *Thymus vulgaris-Bromus erectus* comm. (typical form) complex. It has a low number of vegetation types which are very specific (*e.g.* with the disjunctly distributed *Ephedra helvetica* in the foreground). 510 m, S, Chambave. Photo: A. Schwabe.

Figure 15. Type A2 in the Valle d'Aosta (example): *Melico ciliatae-Kochietum prostratae* complex (see Table 3) with the lowest average values of vegetation types (typical shrubs *e.g. Juniperus communis, Celtis australis*). The complex is restricted to regions with an annual precipitation of about 500 mm. 800 m m.s.l., S, Roisan. Photo: A. Schwabe.

Figure 16. Type A3b in the Valle d'Aosta (example): *Onosmo cinerascentis-Koelerietum vallesianae* complex, montane form with higher average values of vegetation types. The montane influence is indicated by the montane-xerophytic dwarf shrub *Juniperus sabina*. In the foreground: vineyards. 1,000 m m.s.l., SSW, Morgex/Villair. Photo: A. Schwabe.

The results on the level of physiotopes delineate a general principle: Every physiotope type has a wide range of habitat licenses, each of them regionally colonized by identical or corresponding phytocoenoses. The number of these vegetation types agrees to a considerable degree if definable physiotope-types within a macroclimate region are compared. The human influence - if it is not extreme - causes as a rule higher values of vegetation types/plot area ("intermediate disturbance hypothesis", see above).

Naturally, apart from a census further aspects like scarcity, stenoecy, vegetation texture and structure (Barkman 1979; Dierschke 1994: 75 ff., 144 ff.) have to be considered. Information about textural and structural parameters are partly deducible: *e.g.* about life-forms (life-form spectra of the vegetation types) and synusia (see Dierschke 1994: 144).

Applied aspects of the sigmasociological approach

These methodical investigations, which to begin with were carried out as basic research, can also be directly applied in practice. For this first results were obtained. Methods of application are:

Area-related assessment of intervention measures when planning e.g. motorways. From the geobotanical side not only conventional maps can be drawn up, but also defined vegetation complexes be depicted, which characterize the complexity of ecosystems and their diversity on a correct scale. The defined vegetation complexes are a suitable basis for an investigation of zoological plot areas (Kratochwil & Schwabe 1993).

Monitoring of environmental change. In the framework of an applied system, the sigmasociological method is well-suited for monitoring: *e.g.* in the environment impact assessment, conservation planning, or environmental protection research to determine qualitative and quantitative changes in endangered biotope complexes. First results for *Nardus stricta*-grassland in the Black Forest have been obtained by Schwabe (1991a).

The changes in vegetation complexes of extensively used *Nardus stricta*-grassland were studied in the time axis (10 years). Successive complex relevés with sample plots of about 2-4 ha/plot area make it possible to depict changes on the trophic level and in the diversity pattern. The result was that definable zones of changement could be differentiated and mapped (*e.g.* by abandonment and by fertilization, zones without any changements). Changes in the inventory of vegetation types caused by abandonment lead to a decrease in the diversity of vegetation types in later phases of succession.

In contrast to the comparison of landscapes by vegetation maps (see *e.g.* Westhus 1980 who used the Shannon-Wiener function when comparing his data to those gathered by Mahn & Schubert 1962), the comparison of complex relevés has the advantage that many phytocoenoses distributed in micro-areas can be included (often with indicator value), which otherwise is not possible for reasons of obtaining a correct scale.

In relation with "Global Change" there are results showing a migration of species and plant communities of the alpine zone (see *e.g.* Gottfried *et al.* 1994; Grabherr *et al.* 1994). Changes of vegetation complexes can be determined for the last 10 years by comparing vegetation complex relevés of Theurillat (1992a) drawn up in the framework of the MAB-project in the Aletsch region (Valais); Theurillat, in preparation.

Gap analysis. Also for a "gap analysis" (analyzing gaps in the representation of vegetation types within protected areas, see *e.g.* Burley 1988) the direction taken is important. It is the aim of the gap analysis that all ecosystem types worth being protected in reserve areas are sufficiently represented.

Outlook

The vegetation-complex is in consequence a reference unit, by means of which in a first attempt the structural and textural composition and the diversity of the spatial arrangements of habitats can be determined. Areas for further investigations can be sampled after typifying. The advantage is that it is possible to analyze a large area, *e.g.* on transects of several hundred kilometres in length. All prior studies have revealed that the diversity of vegetation types in complex ecosystems ("diversity pattern") is a synthetic characteristic which has integrant qualities. For complete ecosystem studies including the animal communities in the soil only few stands can be analyzed (in the "Solling" and "Göttinger Wald" project *e.g.* two forests, see Schaefer in this volume).

Like in systematics the diversity on the coenotic level is not unlimited, however, there is a graded similarity and the possibility to form types. It is the final aim to typify the diversity of ecosystems in such a way that general laws ("Allsätze", see *e.g.* Mohr 1970) may be formulated. To achieve this a lot more intensive research has to be done.

Acknowledgements

Parts of this work were supported by the "Deutsche Forschungsgemeinschaft".

References

Ansseau, C. & Grandtner, M.M. 1990. Symphytosociologie du paysage végétal. Phytocoenol. 19: 109-122.

Asmus, U. 1987. Die Vegetation der Fließgewässerränder im Einzugsgebiet der Regnitz. Hoppea 45: 23-276.

Balcerkiewicz, St. & Wojterska, M. 1978. Sigmaassoziationen in der Hohen Tatra. pp. 161-173. In: R. Tüxen (ed), Ber. Int. Sympos. Int. Ver. Veg.kde. 1977. Cramer, Vaduz.

Barkman, J.J. 1973. Synusial approaches to classification. pp. 437-491. In: Whittaker, R.H. (ed), Handbook of Vegetation Science 5. The Hague.

Barkman, J.J. 1979. The investigation of vegetation texture and structure. pp. 123-160. In: Werger, M.J.A. (ed), The Study of Vegetation. Junk, The Hague.

Béguin, C., Grandtner, M. & Gervais, C. 1994. Analyse symphytosociologique de la végétation littorale du Saint-Laurent près de Cap-Rouge, Québec. Phytocoenol. 24: 27-51.

Blab, J., Klein, M. & Ssymank, A. 1995. Biodiversität und ihre Bedeutung in der Naturschutzarbeit. Natur und Landschaft 70 (1): 11-18.

Bond, W.J. 1993. Keystone species. pp. 237-253. In: Schulze, E.-D. & Mooney, H.A. (eds), Biodiversity and Ecosystem Function. Ecolog. Studies 99. Springer, Berlin.
Braun-Blanquet, J. 1961. Die inneralpine Trockenvegetation. Fischer, Stuttgart.
Buchwald, K. 1995. Landschaftsökologie - Landschaft als System. pp. 160-178. In: Steubing *et al.* (eds), Natur- und Umweltschutz - Ökologische Grundlagen, Methoden, Umsetzung. Fischer, Jena.
Burley, F.W. 1988. Monitoring biological diversity for setting priorities in conservation. pp. 227-230. In: Wilson, E.O. (ed), Biodiversity. Washington.
Deil, U. 1995. Vegetation und rezenter Landschaftswandel im Campo de Gibraltar (Südwestspanien) und im Tangerois (Nordwestmarokko). Geoökodynamik 16: 109-136. Bensheim.
Deil, U. in press. Vegetation landscapes in Southern Spain and Northern Morocco - an ethnogeobotanical approach. Fitosociologica 29. Pavia.
Dierßen, K. 1979. A classification of community complexes in mires by phytosociological methods. Proc. Int. Symp. Classif. of Peat and Peatlands, Hyytiälä, Finland: 33-41. Helsinki.
Dierßen, K. 1990. Einführung in die Pflanzensoziologie, Vegetationskunde. Wiss. Buchgemeinsch., Darmstadt.
Dierschke, H. 1972. Zur Aufnahme und Darstellung phänologischer Erscheinungen in Pflanzengesellschaften. pp. 305-325. In: Van der Maarel, E. & Tüxen, R. (eds), Grundfragen und Methoden in der Pflanzensoziologie. Ber. Int. Sympos. Int. Ver. Veg.kde. Rinteln 1970. Cramer, Lehre.
Dierschke, H. 1994. Pflanzensoziologie. Ulmer, Stuttgart.
Ehrlich, P.R. 1993. Biodiversity and ecosystem function: need we know more? pp. VII-XI. In: Schulze, E.-D. & Mooney, H.A. (eds), Biodiversity and Ecosystem Function. Ecolog. Studies 99. Springer, Berlin.
Ellenberg, H., Meyer, R. & Schauermann, J. 1986. Ökosystemforschung; Ergebnisse des Sollingprojektes 1966-1986. Ulmer, Stuttgart.
Elton, C. 1933. The Ecology of Animals. Methuen, London.
Feeny, P. 1976. Plant apparency and chemical defense. Rec. Adv. Phytochem. 10: 1-40.
Frankel, O.H., Brown, A.H.D. & Burdon, J.J. 1995. The Conservation of Plant Biodiversity. Cambridge Univ. Press, Cambridge.
Géhu, J.-M. 1991. L'analyse symphytosociologique et géosymphytosociologique de l'espace. Théorie et méthodologie. pp. 11-46. In: J.-M. Géhu (ed), Coll. Phytosoc. 17. Cramer, Berlin, Stuttgart.
Gillet, F. Les phytocénoses forestières du Jura nordoccidental. Essai de phytosociologie intégrée. Thèse, Université de Besan≤on.
Gillet, F. & Gallandat, J.-D. 1996. Integrated synusial phytosociology: some notes on a new, multiscalar approach to vegetation analysis. Journal Veg. Science 7: 13-18.
Gillet, F., de Foucault, B. & Julve, P. 1991. La phytosociologie synusiale intégrée: objets et concepts. Candollea 46: 315-340.
Goetze, D. 1996. Zur Typisierung von Landschaftsausschnitten in Bachtälern des Oden- und Nordschwarzwaldes auf der Grundlage von Vegetationskomplexen. Braunschweiger Arbeiten 4: 259-268.
Goetze, D. & Schwabe, A. 1997. Levels of biodiversity on different scales in space. Comparative landscape-ecological investigations in Southwest Germany. IAVS Symposium Ceské Budejovice. Acad. Science Czech Republic: 38-39. Pruhonice u Prahy.
Gottfried, M., Pauli, H. & Grabherr, G. 1994. Die Alpen im "Treibhaus": Nachweise für das erwärmungsbedingte Höhersteigen der alpinen und nivalen Vegetation. Verein zum Schutz der Bergwelt e.V: 13-27. München.
Grabherr, G. 1994. Biodiversität und landschaftliche Vielfalt Österreichs. pp. 23-56. In: Morawetz, W. (ed), Ökologische Grundwerte in Österreich. Österr. Akad. Wiss., Wien.
Grabherr, G., Gottfried, M. & Pauli, H. 1994. Climate effects on mountain plants. Nature 369: 448.
Haber, W. 1979. Theoretische Anmerkungen zur "ökologischen Planung". Verh. Ges. Ökol. (Münster 1978) 7: 19-30.
Haeupler, H. 1982. Evenness als Ausdruck der Vielfalt in der Vegetation. Untersuchungen zum Diversitätsbegriff. Diss. Bot. 65. Cramer, Vaduz.
Haeupler, H. 1995. Diversität. In: Kuttler, W. (ed), Handbuch zur Ökologie. 2nd ed. Analytica-Verlag, Berlin.
Haeupler, H. in this volume. Elements of biodiversity in today's nature conservation discussion - from a geobotanical viewpoint.
Harper, J.L. & Hawksworth, D.L. 1996. Preface. pp. 5-12. In: Hawksworth, D.L. (ed), Biodiversity. Measurement and Estimation. Chapman & Hall, London.
Hawksworth, D.L. (ed) 1996. Biodiversity. Measurement and Estimation. Chapman & Hall, London.
Heywood, V.H. & Watson, R.T. 1995. Global Biodiversity Assessment. Cambridge Univ. Press, Cambridge.

Huston, M.A. 1994. Biological Diversity. Cambridge Univ. Press, Cambridge.
Kinzelbach, R.K. 1989. Ökologie, Naturschutz, Umweltschutz. - Dimensionen der modernen Biologie 6. Wiss. Buchges., Darmstadt.
Köppler, D. 1995. Vegetationskomplexe von Steppenheide-Physiotopen im Juragebirge. Diss. Bot. 249: 1-228. Cramer, Stuttgart, Berlin.
Köppler, D. & Schwabe, A. 1996. Typisierung und landschaftsökologische Gliederung S- und W-exponierter Jura-"Steppenheiden" mit Hilfe von Vegetationskomplexen. Ber. Reinhold-Tüxen-Ges. 8: 159-192.
Körner, Ch. 1993. Scaling from species to vegetation: the usefulness of functional groups. pp. 117-140. In: Schulze, E.-D. & Mooney, H.A. (eds), Biodiversity and Ecosystem Function. Ecolog. Studies 99. Springer, Berlin.
Kratochwil, A. & Schwabe, A. 1993. Biozönologisch-landschaftsökologische Bestandsaufnahme und Bewertung bei der Umweltverträglichkeitsprüfung (UVS) unter Berücksichtigung von Tiergemeinschaften, Pflanzengesellschaften und Vegetationskomplexen. Schr.-Reihe Forschung, Straßenbau und Straßenverkehrstechnik 636: 63-84. Bonn-Bad Godesberg.
Lawton, J.H. & Brown, V.K. 1993. Redundancy in ecosystems. pp. 255-270. In: Schulze, E.-D. & Mooney, H.A. (eds), Biodiversity and Ecosystem Function. Ecolog. Studies 99. Springer, Berlin.
Leser, 1996. Landschaftsökologie. 4th ed. Ulmer, Stuttgart.
Luder, P. 1981. Die landschaftsökologische Diversität. Begriffsdiskussion und Versuch der empirischen Kennzeichnung. Angew. Botanik 55: 321-329.
Lüttge, U. 1991. Clusia. Morphogenetische, physiologische und biochemische Strategien von Baumwürgern im tropischen Wald. Naturwiss. 7: 49-58.
Mahn, *e.g.* & Schubert, R. 1962. Vegetationskundliche Untersuchungen in der mitteldeutschen Ackerlandschaft VI. Wiss. Z. Martin-Luther-Univ. Halle. Math.-Naturwiss. Klasse 11: 765-816.
Matuszkiewicz, W. & Plit, J. 1985. Versuch einer typologischen und regionalen Landschaftsgliederung auf Grund der Karte der potentiellen natürlichen Vegetation. Phytocoenol. 13 (2): 161-180.
Miyawaki, A. 1978. Sigmaassoziationen in Mittel- und Süd-Japan. pp. 241-263. In: Tüxen, R. (ed), Ber. Int. Sympos. Int. Ver.Veg.kde. 1977. Cramer, Vaduz.
Miyawaki, A. 1980. Versuch der Darstellung eines japanischen Landschaftssystems nach Sigmaassoziationen am Beispiel des Sagami-Flußgebietes in der Kanto-Ebene, Mittel-Honshu. pp. 399-407. In: Wilmanns, O. & Tüxen, R. (eds), Ber. Int. Sympos. Int. Ver. Veg.kde. 1979. Vaduz.
Mohr, H. 1970. Biologie als quantitative Wissenschaft. Naturwiss. Rundschau 7/161: 779-785.
Mühlenberg, M. 1993. Freilandökologie. 3rd ed. Quelle & Meyer, Heidelberg, Wiesbaden.
Neef, E. 1981. Das Gesicht der Erde. 5th ed. H. Deutsch, Zürich, Frankfurt a.M.
Odum, E.P. 1971. Fundamentals of Ecology. Saunders Comp., London, Philadelphia, Toronto.
Odum, E.P. 1983. Grundlagen der Ökologie. Thieme, Stuttgart, New York.
Pignatti, S. 1978. Zur Methodik der Aufnahme von Gesellschaftskomplexen. pp. 27-38. In: Tüxen, R. (ed), Ber. Int. Sympos. Int. Ver. Veg.kde. 1977. Vaduz.
Pignatti, S. 1981. Carta dei complessi di vegetazione di Cortina d'Ampezzo. Consiglio Nazionale delle Ricerche AQ/ 1/ 189. Roma. 39 pp. + map.
Remmert, H. 1984. Ökologie. Ein Lehrbuch. 3rd ed. Springer, Berlin.
Rivas-Martínez, S. 1976. Sinfitosociologia, una nueva metodologia para el estudio del paisaje vegetal. An. Inst. Bot. A. J. Cavanilles 33: 179-188.
Rosenzweig, M.L. 1995. Species Diversity in Space and Time. Cambridge Univ. Press, Cambridge.
Schaefer, M. 1991. Animals in European temperate deciduous forests. pp. 503-525. In: Röhrig, E. & Ulrich, B. (eds), Temperate Deciduous Forests (Ecosystems of the World). Elsevier, Amsterdam.
Schaefer, M. 1992. Ökologie. Wörterbücher der Biologie. 3rd ed. Fischer, Jena.
Schaefer, M. in this volume. The diversity of the fauna of two beech forests: some thoughts about possible mechanisms causing the observed patterns.
Schuhwerk, F. 1986. Kryptogamengemeinschaften in Waldassoziationen - ein methodischer Vorschlag zur Synthese. Phytocoenol. 14 (1): 79-108.
Schulze, E.-D. & Mooney, H.A. (eds) 1993. Biodiversity and Ecosystem Function. Ecolog. Studies 99: 1-510.
Schwabe, A. 1987. Fluß- und bachbegleitende Pflanzengesellschaften und Vegetationskomplexe im Schwarzwald. Diss. Bot. 102: 1-368. Cramer, Berlin.
Schwabe, A. 1988. Erfassung von Kompartimentierungsmustern mit Hilfe von Vegetationskomplexen und ihre Bedeutung für zoozönologische Untersuchungen. Mitt. Bad. Landesver. Naturk. u. Natursch. 14: 621-630.
Schwabe, A. 1989. Vegetation complexes of flowing-water habitats and their importance for the differentiation of landscape units. Landscape Ecology 2: 237-253.

Schwabe, A. 1990. Stand und Perspektiven der Vegetationskomplex-Forschung. Ber. Reinhold-Tüxen-Ges. 2: 45-60.
Schwabe, A. 1991 a. A method for the analysis of temporal changes in vegetation pattern at the landscape level. Vegetatio 95: 1-19.
Schwabe, A. 1991 b. Perspectives of vegetation complex research and bibliographic review of vegetation complexes in vegetation science and landscape ecology. Excerpta Botanica 28 (Sect. B): 223-243.
Schwabe, A. 1991c. Vegetation complexes can be used to differentiate landscape units. Coll. Phytosoc. Versailles: Phytosociologie et Paysages: 261-279. Stuttgart.
Schwabe, A. 1995. *Kochia prostrata* (L.) Schrader -reiche Pflanzengesellschaften und Vegetationskomplexe unter besonderer Berücksichtigung des Aostatales. Carolinea 53 (Festband Oberdorfer): 83-98.
Schwabe, A. in press. Sigmachorology as a subject of phytosociological research: a review. Phytocoenologia.
Schwabe, A., Köppler, D. & Kratochwil, A. 1992. Vegetationskomplexe als Elemente einer landschaftsökologisch-biozönologischen Gliederung, gezeigt am Beispiel von Fels- und Moränen-Ökosystemen. Ber. Reinhold-Tüxen-Ges. 4: 135-145.
Schwabe, A. & Kratochwil, A. 1994. Gelten die biozönotischen Grundprinzipien auch für die landschaftsökologische Dimension? Vegetationskomplexe inneralpiner Trockengebiete als Fallbeispiele. Phytocoenol. 24: 1-22.
Schwabe, A. & Mann, P. 1990. Eine Methode zur Beschreibung und Typisierung von Vogelhabitaten, gezeigt am Beispiel der Zippammer (*Emberiza cia*). Ökologie der Vögel (Ecology of birds) 12: 127-157.
Schwöppe, W. & Thannheiser, D. 1995. Landschaftsökologische Betrachtungen im Senegal-Delta - Planungsvorschläge für nachhaltige Landnutzung. Hamburger Geogr. Studien 47: 27-44.
Solbrig, O.T. 1991. Biodiversity. Scientific Issues and Collaborative Research Proposals. UNESCO Paris.
Stearns, S.C. *et al.* 1990. Biodiversity. pp. 46-74. In: Schweizerischer Wissenschaftsrat (ed), Forschungspolitische Früherkennung. Technologien zur Erhaltung der Biologischen Vielfalt. Bern.
Stocker, O. 1979 a. Ökologie und Soziologie in erkenntnistheoretischer und empirischer Sicht. Phytocoenologia (Festband Tüxen) 6: 1-14.
Stocker, O. 1979 b. Ökologie als existentiales Problem im Viererschema der biologischen Wissenschaften. Flora 168: 13-52.
Strasburger, E. (new edition by Sitte, P., Ziegler, H., Ehrendorfer, F. & Bresinsky, A.) 1998. Lehrbuch der Botanik für Hochschulen. 34th ed. Gustav Fischer Verlag, Stuttgart, Jena, New York.
Templeton, A.R. 1996. Biodiversity at the molecular genetic level: experiences from disparate macroorganisms. pp. 59-64. In: Hawksworth, D.L. (ed), Biodiversity. Measurement and Estimation. Chapman & Hall, London.
Thannheiser, D. 1988. Eine landschaftsökologische Studie bei Cambridge Bay, Victoria Island, N.W.T., Canada. Mitt. Geogr. Ges. Hamburg 78: 1-51.
Thannheiser, D. 1992. Landschaftsökologische Studien in der kanadischen Arktis. Naturschutzforum 5/6: 201-217.
Theurillat, J.-P. 1992 a. Etude et cartographie du paysage végétal (Symphytocénologie) dans la région d'Aletsch (Valais, Suisse). Beitr. Geobot. Landesaufn. Schweiz 68: 1-384. Vol. I, II. Teufen.
Theurillat, J.-P. 1992 b. Abgrenzungen von Vegetationskomplexen bei komplizierten Reliefverhältnissen, gezeigt an Beispielen aus dem Aletschgebiet (Wallis, Schweiz). Ber. Reinhold-Tüxen-Ges. 4: 147-166.
Thienemann, A. 1920. Die Grundlagen der Biocoenotik und Monards faunistische Prinzipien. Festschrift für Zschokke 4: 1-14. Basel.
Thienemann, A. 1956. Leben und Umwelt. Vom Gesamthaushalt der Natur. Rowohlt, Hamburg.
Tischler, W. 1993. Einführung in die Ökologie. 4th ed. Fischer, Stuttgart.
Tüxen, R. 1979. Sigmeten und Geosigmeten, ihre Ordnung und ihre Bedeutung für Wissenschaft, Naturschutz und Planung. Biogeographica 16 (Vol. in Honour of J. Schmithüsen): 79-92.
Ulrich, B. 1993. Prozeßhierarchie in Waldökosystemen. Biologie in unserer Zeit 23: 322-329.
Van der Maarel, E. 1988. Species diversity in plant communities in relation to structure and dynamics. pp. 1-14. In: During H.J. *et al.* (eds), Diversity and Pattern in Plant Communities. The Hague.
Vevle, O. 1988. Synsociological treatment of vegetation complexes in nature reserves of salt marsches in Norway. Coll. Phytosoc. 15: 275-390.
Walter, H. & Lieth, H. 1960 ff. Klimadiagramm-Weltatlas. Fischer, Jena.
Weißbecker, M. 1993. Fließgewässermakrophyten, bachbegleitende Pflanzengesellschaften und Vegetationskomplexe im Odenwald - eine Fließgewässertypologie. Schriftenreihe Hess. Landesanst. für Umwelt 150: 1-156.

Westhus, W. 1980. Die Pflanzengesellschaften der Umgebung von Friedberg (Kr. Hettstedt) und Wanzleben während des Zeitraumes 1978/79 und ihr Vergleich mit Untersuchungsergebnissen von 1958/59 bzw. 1961/62. Diplomarb. Mskr. Martin-Luther-Univ. Halle.
Whittaker, R.H. 1975. Communities and Ecosystems. 2nd ed. New York, London.
Whittaker 1977. Evolution of species diversity in land communities. Evolutionary Biology 10: 1-67.
Whittaker, R.H.S., Levin, A. & Root, R.B. 1973. Niche, habitat and ecotope. Amer. Nat. 107: 321-338.
Wilmanns, O. 1993. Ökologische Pflanzensoziologie. 5th ed. Heidelberg.
Wilmanns, O. & Tüxen, R. 1978. Sigmaassoziationen des Kaiserstühler Rebgeländes vor und nach Großflurbereinigungen. pp. 287-302. In: Tüxen, R. (ed), Ber. Int. Sympos. Int. Ver. Veg.kde. Rinteln 1977. Vaduz.
Wilson, E.O. 1988. Biodiversity. National Acad. Press, Washington.
World Conservation Monitoring Centre (ed) 1992. Global Biodiversity. Chapman & Hall, London.
Zwölfer, H. & Völkl, W. 1993. Artenvielfalt und Evolution. Biologie in unserer Zeit 23: 308-315.

Nomenclature according to:
Binz, A. & Heitz, C. 1990. Schul- und Exkursionsflora für die Schweiz. 19th ed. Schwabe & Co., Basel.

DIVERSITY OF PASTURE-WOODLANDS OF NORTH-WESTERN GERMANY

RICHARD POTT

Institute of Geobotany, University of Hannover, Nienburger Str. 17, D-30167 Hannover, Germany

Key words: Ancient woodland, biotope protection, community diversity, extensively used woodland, primeval forests, species diversity, woodland structures

Abstract

Prehistoric and historic man has not only turned the original landscape into a cultural one (by clearances, land reclamations, and specific cultivation procedures); he has also indirectly influenced the site-differentiating and vegetation-dynamic processes by acting on the natural environmental conditions, mainly on the soil to a small or large degree. Today's distribution pattern of habitat factors probably still largely corresponds to that of former times, however, in individual cases, there may be considerable anthropogenic deviations, so that the previous and the current potential natural vegetation in the same habitats cannot in every case be compared. When investigating this problem, the respective degree of reversibility of the anthropogenic habitat modifications plays a decisive part; it depends on the entirety of human influence with its present and former utilizations, as well as on the habitat-specific regenerative power.

This is especially true for the pasture-woodlands, relics of which have been preserved up to this day in some national nature reserves in north-western Germany.

Originally a mainly closed woodland cover could be found in Central Europe; in prehistoric times and in the Middle Ages, this was being changed by man into a mosaic of highly varying, half-natural and half-open vegetation structures, like meadows, hedges, shrubs, heathlands interspersed with juniper, and a large number of different woodland types. The increase in structural variety also led to a distinct enrichment of flora and vegetation. Today, these forests are of extraordinary importance as biological reservation landscapes; they serve as refuges for numerous plant and animal species. In our landscape, some witnesses of the old utilization forms can still be found; however, since their number is continually decreasing, it can be foreseen when they will have disappeared, too. The exemplary protection and preservation of these last witnesses is therefore to be recommended, using all possibilities of and all the knowledge gathered in the fields of landscape and forest planning. Corresponding programmes have to be developed by an ecological management, under the guidance of qualified biologists and ecologists.

A. Kratochwil (ed.), Biodiversity in ecosystems, 107-132

Introduction

Biodiversity has today, several years after the United Nations Conference for Environment and Development, held at Rio de Janeiro in 1992, become a modern catchphrase in the discussion about the state and the treatment of our landscape. Original landscape, natural landscape, cultural landscape, and industrial landscape are terms deduced from the different current utilization forms of special landscape parts. In this context, however, it is often forgotten that the allegedly natural landscapes with their characteristic natural features have been subject to constant changes for millenniums. For many people, mudflats and islands, lakes, rivulets, bogs, heathlands, and forests, the moderately used grasslands and meadows with their hedges and shrubs, as well as the pasture-woodlands, are representatives of intact landscapes, since here often a high biotope variety can be found in a very small area. Varied and differentiated landscape parts are as a rule also valuable habitats, rich in species and with a high biodiversity. This essay will give some insights into the variety within the pasture-woodlands; this is all the more important as numerous habitats of wild plants and animals are today highly endangered in their existence.

The original habitats have nowadays often become consumer goods for man, and this fact characterizes large areas of today's landscapes. However, the historical development of the cultural landscape also shows that the anthropogenic influence on vegetation cannot solely be regarded from the point of view of modern man. Man has not only destroyed nature, but has also acted on vegetation and landscape in a highly enriching and differentiating way. As to the influence of man, we have to distinguish between two subsequent processes: the period of enrichment of vegetation and the period of impoverishment of vegetation.

The process of enrichment of vegetation starts with the first interventions of the Neolithic peasant in the original forest landscape and essentially covers the long period of rural common land cultivation. The number of plant communities and the variety of flora and vegetation largely increased, compared to the natural landscape which is only differentiated to a very small extent. When cultivating the land, man turned the former, nearly closed hardwood landscape into today's cultural landscape, a mainly open and intensively used cultivated landscape dominated at present by anthropogenic forest stands and plant communities.

The numerous, today often outdated rural cultivation forms applied to fields and forests are usually given as reason for the anthropogenic changes in vegetation and landscapes. As different environmental factors, they caused specific vegetation formations (*e.g.* coppices, pasture-woodlands, winning of leaf litter, slash and burn-cultivation etc.). The same is true for the agricultural cultivation forms (*e.g.* ley farming, *i.e.* rotational cultivation systems with varying rotation times). Greenland utilizations (*e.g.* meadow and pasture management, pastures, heathland stretches, mesoxerophytic meadows, meadows used for rotational grazing, as well as the establishment of litter meadows with reed utilization etc. on moist biotopes) also created new habitats with corresponding plant and animal species.

As a consequence of these old, often not regulated, and exploiting management forms, mainly the half-natural biotope types with their respective vegetation forms developed. As long as these forms were retained, the cultivated landscapes with their special vegetation units were not endangered, since the interventions of man (and of animals) constituted specific environmental factors, which were a prerequisite for the existence of these landscapes.

In the natural landscapes, the more subtle habitat factor differences in the natural location mosaic are levelled out under a closed woodland cover and are therefore not effective. This is true for the local climate of the forest, which has an assimilating effect, as well as for the slighter differences in the soil mosaic. However, after deforestation in half-natural secondary biotopes which are poor in coppice or do not possess any small wood at all, these differences fully take effect. This applies both to the open land climate characterized by contrasts in small areas and to the fine-mosaic of the soil. Even if one forest association is always used in the same way, several anthropogenic secondary communities may develop from this association, and from the original varying forest landscapes specific, partly even characteristic cultural landscapes with the corresponding vegetation inventory have arisen. This process of diversification of the landscapes entailing an enrichment and a differentiation of vegetation was followed by the period of impoverishment in vegetation and dedifferentiation, that is to say, a recessive development began. This development was due to an intensification of agriculture and of the technical and civilizatory progress; the process coincided with the period of agricultural intensive management. It started in modern times, when, mainly at the beginning of the 19th century, the marks and borderlands were compartmented and mineral fertilizer was introduced, and it has seriously intensified in the last decades as a consequence of the unchecked technical and civilizatory progress. Pasture-woodlands and pasture-woodland landscapes are thus relic structures from former times with a relatively high diversity of vegetation and fauna, compared to intensively used landscapes. The biodiversity of such extensively cultivated landscapes can be shown for different complexity levels: many half-natural small ecosystems with corresponding structures and functions, which depend on the anthropo-zoogenic utilization, enrich these extensive landscapes.

Near-natural old forests and pasture-woodlands

In Central Europe, there are only few primeval forests left, which come up to the original definition of a primeval forest; they have, as it can be proven, not been subject to direct anthropogenic interventions (see, among others, Zukrigl *et al.* 1963; Prusa 1985; Mayer *et al.* 1987; Leibundgut 1993; Pott 1993). These forests can only be found in the upper parts of the low mountain range and in the alpine region; there they are restricted to steep or very stony, inaccessible slopes. The more easily approachable and utilizable forests in the flat parts of the low mountain range and those of the lowlands have long since been anthropogenically influenced in many different ways (Fig. 1).

The forest vegetation of former times is the result of a natural change in vegetation sequences caused by climate and succession. However, already at the oak-mixed forest period in the Atlantic Period (ca. 6000-3200 B.C.), the natural development processes were disturbed and partly even prevented by the settlements of Neolithic man. Man did not interfere in a static vegetation condition of the woodland landscapes of that time, but in a dynamic process, which was far from being finished. Fir (*Abies alba*), beech (*Fagus sylvatica*), and hornbeam (*Carpinus betulus*), *e.g.*, had not yet completely established themselves in the wood species inventory; thus the formation of these types to mixed coniferous forests, to beech-, beech-mixed- and oak-hornbeam forests, which today belong in many regions to the dominating potential woodland communities, had not even started. Since the more recent Atlantic Period (*e.g.* from 4500 B.C. on the loess soils in south and Central Europe and from 3200 B.C. in the geest regions in north-western Europe), the natural development of vegetation interlocked with the influence of man; ever since, there has been no covering original natural forest landscape with a corresponding vegetation.

Figure 1. Natural primeval forests show a high diversity of vegetation and fauna. The large structural variety in the optimum phase of a natural forest offers many microhabitats (Langenbruch in the Sauerland, 1986).

The natural landscape, primarily a forest landscape, was gradually being turned into a - regionally highly varying - cultural landscape; this process was influenced by occasional setbacks and periods of settlement in certain areas. Vegetation and landscape formations of different bygone eras and of today have to varying degrees been influenced and thus characterized by man. This is, as already mentioned above, especially true for the forests in the cultural landscape. Many previous, extensively used woodland systems, like the pasture-woodlands and the lopping forests, are, due to their relic character, nowadays important research objects to elucidate and interpret such cultivation forms, as well as their effects on the vegetation and landscape formation in large areas of Central Europe and in other regions. Numerous abandoned and recent community forests had usually developed from common property to hunting districts under the ban of the corresponding sovereign, due to the mark-like constitutions of the Middle Ages. The stand structures of these forests show even today still some characteristic features of wood-pasture and its side utilizations (Fig. 2 and 3). In common parlance, they are designated as "primeval forests" because of the crooked forms of their trees and their selective structure, however, quite the contrary is the case: these forests are altogether characterized by extensive former utilization, in some parts, today again a natural succession sequence can be found.

Figure 2. "Hasbruch" on the "Delmenhorster Geest" ("Delmenhorst geest"). The oak-hornbeam forest, used for centuries as combined pasture-woodland and hunting district, shows even today still traces of the former utilizations (which had been continually documented since 1231).

Owing to their present stand structure and their physiognomy, the old, extensively used forests often convey the impression of being natural and original. Despite and because of their former extensive utilization, they often display nowadays again a natural succession sequence, and that makes them so fascinating. However there are also clearly definable criteria concerning the biological and site diversity of such forest areas, which possess in today's intensively used cultural landscapes, like islands, a higher biotope and species diversity, and offer, as remnants of natural, half-natural, or near-natural vegetation units, suitable habitats for many domestic plants and animals.

Figure 3. Stand composition and general structure of a hunting forest reserve: the highest mast oaks tower above the forest aspect; the inferior tree stratum mainly consists of old pollarded hornbeams, which were subject to lopping (in order to obtain leaf hay) or simply used to gain pollard. This form of forest management had several advantages. The same wood stands could be utilized both as pasture-woodland and as lopping forest, since at the usual lopping heights of 2-2,50 m the twig shots were no longer threatened by browsing cattle. Thus it was not necessary any more to stop grazing for several years, and the pasture-woodland could be used on three "floors": at the bottom as meadow, at a height of 2-2,50 m as producer of leaf hay, and at the treetops as mast producer (after Burrichter 1984 and Pott 1993).

They are of high cultural and historical importance, too. In many recent articles, the relations between habitat, vegetation, structure, and dynamics in these forests have been described (see *e.g.* Falinski 1986; Koop 1989, 1991; Pott 1981, 1988a; Pott & Hüppe 1991). The relevance of such forests (forest reserves, lopping forests, pasture-woodlands, coppiced woods) for the species and biotope protection and preservation, as well as for biocoenological research (concerning *e.g.* the habitat claims of animal species) is nowadays reflected in the manifold efforts to protect and conserve nature and ecosystems (see Federal law on the protection and preservation of nature for Germany, § 20c). The European Science Foundation (ESF, Brussels, Strasbourg) has established a research group (Forest Ecosystem Research Network, FERN) to scientifically investigate such forest types (see *e.g.* Salbitano 1988; Pott 1988b; Schuler 1988; Emanuelson 1988; Peterken 1988; Watkins 1988). Another research group, Ancient Woodland, provides data on the history of woodlands and forestry, on vegetation and landscape development; moreover, a comparison of the prehistoric and historic land utilization in different European countries has been presented (see Teller 1990). There is no concrete, linguistically correct translation for the term "ancient woodland" (used to designate forest plots with a provable stand continuity before the fixed year 1600 A.C.; see Peterken 1981; Peterken & Game 1981, 1984) into the German language; it would have to be equated with the terms "altes Waldland" ("ancient woodland"), "aus alter Zeit stammendes Waldland" ("woodland dating from old times"), "altertümliche Wälder" ("ancient forests"), or "ehemalige Wälder" ("former forests"). The expressions "historisch alte Wälder" ("historically ancient forests") and "historisch alte Waldflächen" ("historically ancient forest areas") constitute a pleonasm and are therefore linguistically not acceptable; "Altwald" ("ancient forest"), "Altwälder" ("ancient forests"), or "altes Waldland" ("ancient woodland") are preferably chosen to denote the traditionally utilized forest areas in the cultural landscape (see Pott 1994, 1996). These forests have, as it can be proven, been continually containing woodland stands for more than 400 years, have old, profound soils with a corresponding vegetation, and can thus be distinguished from the more recent woodlands and forest plantations (= recent woodland; see Rackham 1976, 1980; Peterken 1981).

Forest-preserving measures as basis for the present existence of near-natural old forests

All up to now described management forms (coppice management, pasture-woodlands etc.) were basically employed in commonly owned areas, *i.e.* in the common marks. A general change in the cultivation of the forests in Central Europe was induced by the division of the land into marks, divisions that mainly started in the last half of the 18th century, and in some areas continued in the 19th century. These common land divisions are the beginning of a more or less regulated forest and woodland management, so that the period of the devastation of the forests was followed by an era of reafforestation.

Even before the division into common marks was effected, in many regions so-called "private woods" existed, which were distinguished from the "cumulative woods". There were a lot of intergrades between those two basic forms, the common mark on the one hand and the private wood on the other. The private woods were largely owned by the sovereigns, the nobility, and the ecclesiastical institutions. There were different ways to put them into private ownership: they were either transferred from mark-constitution structures or singled out from common marks and deer parks; some of them were anyhow situated on hereditary private property. The cumulative woods are old forest areas on mark ground, which as a rule were in the possession of a closed entitled party, and in which the respective sovereigns had a share.

Most of the forest devastations in the Middle Ages were so disastrous that already in the late Middle Ages many institutions or sovereigns, respectively, saw themselves compelled to make the utilization of the woodlands in some common land regions within the territories under their jurisdiction their own, exclusive right. Apart from their private property, they declared expansive areas to be protected regions (not altogether for unselfish reasons: they often wanted to use them as hunting grounds). This way, the extensive utilization could be moderated and the impending devastation of the concerned forests and land stretches be avoided. In north Germany, a region poor in forests, examples for such hunting districts under a ban were *e.g.* the "Bentheimer Wald" ("Bentheim Forest"), the "Neuenburger Urwald" ("Neuenburg primeval forest") near Varel i.O., the "Hasbruch" near Delmenhorst, and the "Baumweg" region north-east of Cloppenburg (Fig. 2-5), the expansive woodlands in the national park Müritz and in the biosphere reserve Schorfheide/Chorin in Brandenburg. Similar and comparable transfers of possessory rights, *i.e.* the transformation of common lands into hunting districts under the ban of the respective sovereign and the entailing utilization restrictions have often been described and documented for all regions in Central Europe (see, among others, Hesmer & Schröder 1963; Pott & Burrichter 1983; Burschel & Huss 1987; Mantel 1990; Pott & Hüppe 1991; Pott 1993, 1996; Härdtle 1995).

Figure 4. Bizarre deformations of oak trunks in the nature preserve "Baumweg" near Ahlhorn i. Oldenburg (taken from Pott & Hüppe 1991).

The owners of the private forests were usually the only ones entitled to used them, however, with certain restrictions, most people were allowed to let their cattle graze in these areas. Already at the beginning of the 12th and 13th century, many sovereigns assumed the usufructuary right for all former common mark forests within the territory under their jurisdiction, and from this time on, hunting, clearing (in some regions particularly the felling of oaks), and mast were forbidden in large areas. Up to modern times, these limitations were especially strict if the hunting interests of the sovereigns were affected; in such areas, near-natural high forests as refuges for game could still be found. The decrees were *e.g.* issued to preserve oaks and beeches for mast purposes, or for the winning of structural timber from oaks, which already began in the Middle

Ages. These woodland stands important for mast were often altered to that effect that finally merely the high trunks of overmature oaks or beeches rose above overmature shrub stands; their natural loss was the end of the forest. The gradual possessory transfers of suitable forest plots from the common march-like constitutions of the Middle Ages into hunting districts under the ban of the corresponding sovereigns often entailed significant limitations of the former mark-cooperative usufructuary rights; so many forests regions were still extensively utilized, but no longer completely overused. The different common mark constitutions were thus responsible for the varying degrees of utilization. Forest marks in which a sovereign had a share were, usually for hunting reasons, aimed at a preservation of the tree stand, whereas in the common marks with their more or less open rough pastures extensive pasture farming was more important than the conservation of the forest.

Figure 5. Forest aspect of the former pasture-woodlands and lopping forests in the "Bentheimer Wald" showing a characteristic species, structural and functional diversity on the once utilized levels. These forest types have today, after abandonment of the old utilization forms, reassumed the appearance of near-natural forests. In common parlance, they are even called primeval forests.

Mainly the hunting districts under a ban show striking similarities as to their structure, similarities which are usually due to their utilization as grazing area for cattle. In the forests, the high mast-bearing trees tower above the other trees; the inferior tree stratum mainly consists of old coppiced and pollarded hornbeams, which were subject to lopping (in order to obtain leaf hay) or simply used to gain pollard (see Fig. 2-5). This form of forest management had several advantages. The same woodland stands could be utilized both as pasture-woodlands and as lopping trees, since at the usual lopping heights of 2-2,50 m the twig shots were no longer threatened by browsing cattle. Thus it was not necessary any more to stop grazing for several years, and the pasture-woodlands could be used on three "floors": at the bottom as meadow, at a height of 2-2,50 m as producer of leaf hay, and at the treetops as mast producer (Fig. 3). This combined form of management is comparable to the modern practice of alternate mowing and grazing; however, the varying forms of utilization were not employed at different times, but on different levels.

Often forests in frontier areas were declared to be hunting districts under a ban; they thus constituted a kind of large, natural buffer zone between territorial interests. The "Hasbruch" and the "Neuenburger Urwald" lie in the boundary marks of the former grand duchy Oldenburg; the "Bentheimer Wald" marks the border between the Bentheim sovereigns and the bishops of Münster; the "Sachsenwald" ("Saxony forest") is located on the frontier between Holstein and Hannover; the "Bialowiecza forest" in Poland lies in the border area to White Russia etc. Some forest regions could only be used to a very slight extent or not at all. Often they belong to the community complex of humid oak-hornbeam forests of the type *Stellario-Carpinetum* (phytosociological nomenclature according to Pott 1995). Owing to their water-logged and gleyed loam soils, they could only comparatively late - *i.e.* in the Middle Ages - be durably cultivated. When the extensive forest and wood utilizations were stopped, these forests quickly recovered from the suffered damages, due to their high regenerative power, and meanwhile have reassumed the appearance of near-natural forests (cf. Fig. 4 and 5).

The relevance of near-natural old forests for the biotope and species protection

From the geobotanical point of view, such forests can serve as basis for vegetation-historical, phytosociological-synecological, and environment-related studies. At the same time, they can be a subject of landscape-ecological and -geographical investigations, since they represent typical examples of different kinds of the historical landscape; their plant cover shows even today still traces of previous utilizations. It is possible to reconstruct vegetation units specific to various cultivation forms of former eras and to interpret the current vegetation accordingly. The outward character of today's vegetation landscapes is very individual, due to the varying former utilization forms, and they possess a high structural and functional diversity.

We have shown that those forests which from the Middle Ages until the beginning of modern times were owned by sovereigns, other noblemen, or monasteries, were often used differently than those which were from the earliest settlements up to the 18th century common land (*i.e.* the property of all peasants of one peasantry or of a village). These commonly owned forests were subject to numerous, often rigorous utilizations; when in the second half of the 18th century the common land cultivation ended, they were either completely deforested or only contained loose shrublands. Stenoecious forest species had therefore better chances of survival in the less intensively used hunting districts under the ban of the corresponding sovereigns than in the forests within commonly owned land (cf. Hesmer & Schroeder 1963; Emanuelson *et al.* 1985; Pott & Hüppe 1991; Wittig 1991). Such "ancient forests" are thus valuable relic sites for stenoecious forest species (Fig. 6-8). When modern forestry was introduced, these species partly spread again into neighbouring woodlands; quite often however they are still limited to the relic sites or their closer environment, respectively. The names of the forests generally reflect the former possessory interests: designations like "Klosterholz" ("Monastery woods"), "Paterholz" ("Padres' woods"), "Nonnenbusch" ("Nuns' shrub"), "Königsforst" ("King's forest"), and "Herrenholz" ("Lord's woods") are remnants of those times. The term "Tiergarten" (deer park) shows that the corresponding woodland served as hunting ground for the sovereign or other members of the nobility. As a rule, the hunting grounds were fenced in, and thus protected from wood-pasture, which was largely responsible for the degradation of the commonly owned forests. Consequently they today count among those forest sites which are particularly rich in species (Dinter 1991). The "Sundern" are an example of ancient forests, too, since they were already at an early stage singled out from the common mark by the noblemen and declared to be private property. The German designation "Sundern" reflects this process: the term was derived from "aussondern", German for "to single out". The "Sundern" thus also belong to the hunting grounds under a ban. They all show a typical, always recurring stand structure, recognizable even long after their abandonment, and very old trees can be found there (Fig. 3 and 9, Table 1).

Figure 6. Primula elatior in the "Hasbruch" in 1927 (photo archives of the Westphalian Museum for Biology, Münster). The primrose is an indicator for moist and, for a short time, wet eutrophic gley soils; nowadays, after drainage of the soils, it has become rare.

In today's cultural landscape, the half-open, highly dynamic pasture-woodlands are traditionally used areas comprising open extensive pastures, sections with shrubs or park-like stretches, loose tree stands, and the actual relic woods, and they therefore constitute important biological reservate landscapes with a great diversity (cf. Fig. 10 and 11). This has not only been demonstrated by the phytosociological-floristic studies of Burrichter *et al.* (1980) and Pott & Hüppe (1991), but also by later, mainly biocoenologically oriented publications (based on the afore-mentioned studies) of the Osnabrück study group of Kratochwil (see *e.g.* Kratochwil 1993; Aßmann & Kratochwil 1995; Kratochwil & Aßmann 1996). The results of the investigations of Aßmann (1991) about the Coleoptera coenoses of different pasture-woodland biotopes are impressing.

Figure 7. Gagea spathacea is a relic of ancient forests. The species has become extremely rare, and can only be found in ancient forests ("Hasbruch", 1989).

The main focus of these studies lies on the influence of historical processes on today's species structure of the Coleoptera, their population structure, and the geographical differentiation of relic species. The pasture-woodlands of the Emsland, extensively utilized in the Middle Ages, include a number of special habitats (synusia) and small ecosystems, which have nearly completely disappeared from the surrounding cultural landscape (cf. summarizing essay by Pott & Hüppe 1991; Pott 1996). These site types contain stands of dead and old wood, which are inhabited by many xylobionts. Since most species of this life-form have differentiated demands on their development site (*e.g.* favourable exposition or bark damages at trees), varying synusia are to be expected in view of the polymorphous growing forms of the mast and solitary trees. The biocoenological investigations of the Osnabrück study group have shown that a number of highly endangered species occurs in these pasture-woodlands, *e.g. Colydium elongatum*, *Harpalus neglectus* and *Harpalus seripes*, or *Elaphrus aurens*, to name but a few (see also Aßmann 1991). Silvicolous animal species are characteristic relics of the former large hunting districts ("Tinner Loh", "Baumweg", "Bentheimer Wald", "Hasbruch", "Neuenburger Wald"); among them count *e.g.* the woodland species *Carabus glabratus*, *C. auronitens*, *C. problematicus*, *C. violaceus* (see Arndt 1989; Hockmann *et al.* 1992; Aßmann 1991, 1994), the ground beetles *Pterostichus metallicus*, *Abax ovalis*, and also the spotted salamander (*Salamandra salamandra*, cf. Feldmann 1981). Particularly worth mentioning are the beetles living in dead wood, like the scarabaeus species *Osmoderma eremitica*, and the giant stag beetle (*Lucanus cervus*), larger populations of which can occasionally still be found *e.g.* in the "Hasbruch" in north Germany.

Figure 8. Oxalis acetosella grows epiphytically on the mosses at the trunk bases of old hornbeams in the "Hasbruch" (1989); an example for diversity in microsites.

Figure 9. The "Frederikeneiche" ("Frederike's oak") in the "Hasbruch" is with its approximately 1600 years one of the oldest still living oaks in Germany ("Hasbruch", 1990).

Table 1. Structural differences in the "Bentheimer Wald" (taken from Pott & Burrichter 1983)

Economically used high forests	Former pasture-woodlands and lopping forests
1. Regular stand structure of trees of the same age	Irregular stand structure of trees of different ages (alternation of clearings and shadowy parts)
2. Only few shrubs	Extensive, irregularly distributed shrub layer
3. Few prickly, thorny, or aromatic shrubs	Mainly prickly, thorny, or aromatic shrubs, among them facies-forming *Ilex aquifolium*
4. Rather evenly distributed herb flora	Irregularly distributed herb flora, the coverage degrees of which highly differ

Figure 10. Park lands which developed due to wood-pasture in the "Borkener Paradies" ("Borken Paradise") / Emsland with zonally arranged vegetation complexes, consisting of pastures, shrubs, and woodland (taken from Pott & Hüppe 1991).

The populations of such woodland-inhabiting species, which were probably being isolated from each other in the course of the forest devastations and destructions in the Middle Ages, are not only outstanding from the faunistic point of view, but also from the evolution-biological one. The stenoecious, silvicolous slugs (*Limax cinereoniger*, *L. tenellus*) are a very good example for this. Not only the genetic variability, but also isolation phenomena of these animal populations can be investigated in the old relic forests. First exciting results about the genetic variability in *Carabus glabratus* and *Abax ovalis* have been documented in the publications of Aßmann (1990a, 1990b) and Aßmann *et al.* (1994). But not only the forest-inhabiting beetles constitute an interesting particularity; the ants (mainly the yellow ant *Lasius flavus*) and their syndynamic interactions with special *Sedo-Scleranthetalia*-communities are studied as to the congruence of the collecting behaviour of these society-forming insects. Moreover, the growth of typical, ant-spreading plant species is compared (Kratochwil 1993). Ants play an important part as architects of typical landscape structures.

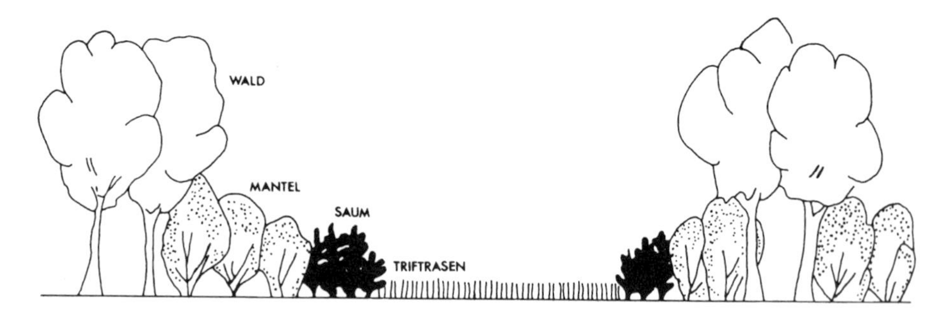

Figure 11. Schematic depiction of the community structure in the pasture-woodland vegetation complex "Borkener Paradies" (taken from Pott & Hüppe 1991).

The studies aim at gaining more detailed knowledge about the interaction of plants and ants (Myrmecochorie) in heath landscapes. The pollinator communities of wild bees in selected typical plant communities of the pasture-woodland mosaic (*e.g. Spergulo-Corynephoretum canescentis, Airetum praecocis, Diantho-Armerietum, Corno-Prunetum*, Fig. 12 and 13) allow fascinating insights into the correlations of plant communities and special animal populations. The migration behaviour of most wild bees is only little pronounced, and as a rule their dispersal ability is relatively low, so that "insulations" can easily occur (cf. also Kratochwil 1983, 1984, 1988; Pott 1996).

The butterfly communities in pasture-woodlands are, due to their preferences for special food plants, bound to certain vegetation units, both in their adult and in their larval stage. Among them comparatively rare species can be found which are specialized in this habitat, *e.g.*, according to Bathke (1994), *Hipparchia semele*, *Lycaena phlaeas*, *Plebejus argus*, *Polymmatus icarus*, *Celastrina argiolus*, and *Hesperia comma*. Like most wild bees, many butterflies show little inclination to migrate, thus also here isolated populations develop. Avifaunistic investigations are still in their initial stage. The results of first studies on stand sizes, habitat choice and habitat utilization of titmice (*Paridae* and *Aegithalidae*) performed by Niemann (1994) in pasture-woodlands promise to lead to highly interesting new findings in future.

The ancient forests, primarily the old hunting districts containing stands of trees which are several hundred years old and old forest soils, are of special importance (see among others Hesmer & Schröder 1963; Ellenberg 1986; Falinski 1986; Hermy

1989; Hermy & Stieperaere 1981; Peterken & Game 1984; Brunet 1992; Korpel 1995; Pott 1996). As already mentioned, many characteristic wood plants - mainly forest-inhabiting grasses, ferns, and special characteristic *Fagetalia* species, as well as mosses, fungi, and lichens - find typical relic sites in these districts, *e.g.* spring bitter vetch (*Lathyrus vernus*), herb Christopher (*Actaea specata*), liverwort (*Hepatica nobilis*), wood anemone (*Anemone nemorosa*), truelove (*Paris quadrifolia*), woodruff (*Galium odoratum*), dog mercury (*Mercurialis perennis*), yellow gagea (*Gagea spathacea*), satinflower (*Stellaria holostea*). Many of these wood plants show special dispersal strategies; the seeds of truelove (*Paris quadrifolia*), *e.g.*, are dispersed by snails, those of *Sanicula europaea* and *Viola reichenbachiana* by elaiosomes and ants.

Figure 12. Small-scale mosaic by ant sandhills in a xerophytic habitat in the "Meppener Kuhweide" ("Meppen cow meadow") (taken from Pott & Hüppe 1991).

Forest grasses like millet grass (*Milium effusum*), wild rye (*Hordelymus europaeus*), brome-grass (*Bromus ramosus*, *Bromus benekenii* and *Melica uniflora*) and fescue grass (*Festuca altissima*) can also be detected in these residual sites. Their occurrence is by no means unusual; it can be explained by the development and the long-lasting, generally moderate utilization of the ancient forest plots. Analyses of pollen from small fens in the surroundings of these forests and from soil profiles give evidence of the immigration times and survival rates of such plants in north-west Europe (cf. among others Isenberg 1979; Aaby 1983; O'Connell 1986; Elerie *et al.* 1993; Pott 1993). Already in the Atlantic Period, the wood plants advanced, together with the deciduous woodlands, from their glacial refuges far towards north, to their current area borders. Ever since, they have been able to survive in the corresponding forest sites. In the course of the post-glacial spreading of the beech and its "companions" into northern direction, many of the partly light-requiring *Fagetalia* species were forced to retreat to the oak-hornbeam forests (*Stellario-Carpinetum*) and other moist forests (sloping or ravine forests of the *Tilio-Acerion*). There they have managed to survive in wooded habitats up to the present day; *e.g.* the higher plants *Stellaria holostea, Carex sylvatica, Circaea lutetiana, Gagea spathacea, Sanicula europaea, Festuca gigantea, Stachys sylvatica*, as well as the mosses *Isothecium myosuroides* and *Thuidium tamariscinum*, and other species listed in the vegetation tables of the forest reserves "Hasbruch", "Bentheimer Wald", "Neuenburger Wald", "Baumweg", and "Tinner Loh", drawn up by Pott & Hüppe (1991). As a rule, these species are missing in secondary forests; on primary sites, the populations of such woodland elements often attain an age of 200-300 years (*e.g. Hepatica nobilis* and *Sanicula europaea*; Inghe & Tamm 1985); this explains why they are so sensitive to forest destructions and site changes.

Figure 13. Vegetation mosaic composed of open sandy stretches, pastures, shrubs, and forest islands in the pasture-woodland "Borkener Paradies" on the Ems (1980). In this region, the former birch-oak forests were replaced by completely new vegetation types, owing to the influence of wood-pasture. An example for anthropo-zoogenic biodiversity!

The ancient forests, however, possess a striking multitude and variety of old ornamental and cultivated plants, which have survived up to now as "wood plants"; among them count *e.g.* the meanwhile naturalized evergreen (*Vinca minor*), the Eurasian-continental leek (*Allium paradoxum*), which originally comes from the Caucasus and Turkmenia, the also naturalized, more submediterranean aquilegia (*Aquilegia vulgaris*), and the east-submediterranean Christmas flower (*Eranthis hiemalis*).

For these plants, the near-natural ancient forests are very important refuges. Often the wood plants are regarded as "indicators" for such forests, which is of course, considering the refuge character of these species, not quite correct. The often only local occurrences of wood plants are due to individual, varying reasons. Each of the pasture-woodlands and hunting districts was subject to partly highly special anthropo-zoogenic utilizations, which accordingly entailed individual successions or stand structures. Thus every forest has its own historical development with all comparabilities, but also all its special features. Indicators for these forests are rather plants which developed as a consequence of traditional grazing and lopping, *e.g.* the increased occurrence of grazing- and trampling-resistant species, like blackthorn (*Prunus spinosa*), hawthorn (*Crataegus monogyna*), whitethorn (*C. oxyacantha*), dogrose (*Rosa canina*), holly (*Ilex aquifolium*). The generally altered structure of these forests is due to the influence of grazing, too (see Table 1). This distinguishes them clearly from the modern economically utilized forests (see Fig. 2-11).

The old forest stretches which were never completely overused or lastingly and vigorously modified and destroyed, often contain as "permanent forest islands" the flora and fauna inventory typical of their region, and are very rich in structures. However, they must cover a certain minimum area: after Zacharias & Brandes (1989, 1990) and Zacharias (1993) a minimum size of about 500 ha is optimal for a near-natural wood stand rich in structures. Recolonizations of secondary forests with wood species typical of the area usually take a very long time: depending on the regenerative power of the wood species, from 350 years (vital *Carpinion* forests; see Falinski 1986) up to 600-800 years (*Quercion roboris* forests growing on acid soils; cf. Peterken 1977; Rackham 1980). Many of the ancient forests which have not been used for forestry for several decades or longer, are characterized by huge, partly dead oak, beech, and hornbeam trees. Those latter serve as habitat for beetle and fungi species living in dead wood, which have, due to a lack of suitable sites, become rare in Europe. Such forests are important as research objects for the common natural heritage, because they possess a great richness in structure and diversity, and an extraordinarily high number of rare plant and animal species.

Pasture-woodlands in north-west Germany

Pasture-woodlands show a relatively high biotope diversity, as their vegetation and physiognomy are characterized both by the natural site conditions and the corresponding grazing intensities and modalities. Since these factors may change within a certain area as well as within a certain period of time, there is no uniform pasture-woodland type, however, comparable features can be ascertained as a consequence of the effects of grazing and pasture selection (cf. Fig. 14 and 15).

According to the differentiated site conditions in the field, each natural forest community constitutes a different starting point for the pasture-woodland vegetation. It represents a frame, within which the varying modifications caused by grazing may occur. In this context, the resistance to browsing and regenerative power of the wood species - which depend either on the site or are species-specific - play, among others, an essential part. The difference *e.g.* between hardwood and coniferous wood is striking. After browsing, the young growth of woodland species can form stool shoots again, whereas the young growth of coniferous wood is, as a rule, not able to do so. Thus the grazed coniferous forests lack, contrary to the utilized hardwoods, the characteristic shrub forms. There are also considerable differences between the individual hardwood communities as to their power of reproduction and their biotope structure, even if they do not take on such exclusive forms as those between hardwood and coniferous wood (see Fig. 16-18).

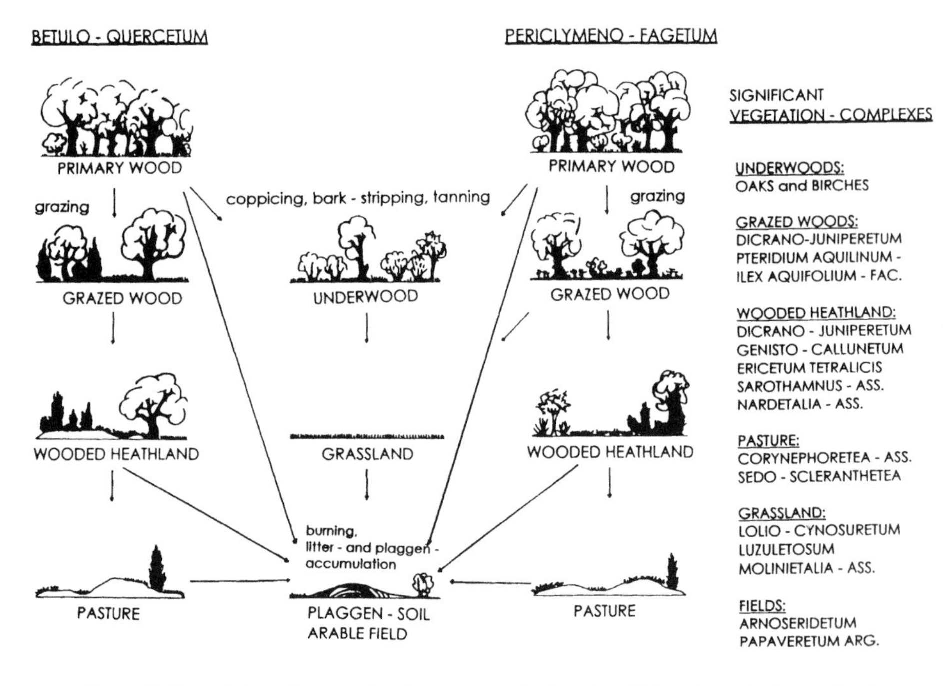

Figure 14. Degradation scheme and anthropo-zoogenic diversity of *Quercion roboris* woodland, due to extensive management in the Pleistocene landscape of NW-Europe.

It is meanwhile generally known that the succession evoked by the grazing cattle leads from a closed forest via loose stands and park-like stages to an open pasture. This is usually due to a continuous degradation series at a lasting grazing intensity. Apart from the individual degradation stages, however, also regeneration complexes may develop in a pasture-woodland, primarily when the grazing intensity decreases and excellent regeneration conditions prevail.

For many pasture-woodlands, such a coexistence of degradation and regeneration complexes is even characteristic. It is just these wood initials, irregularly distributed in the field, which give, together with larger forest relics and individual trees, the areas their special appearance of scenic park lands (see Fig. 10, 11, 12, and 14). The individual complexes show a zonal structure. They consist of regular arrangements of pasture, herbaceous fringe, forest edges (as shrubs), and forest, individual trees, or clusters of trees, respectively; the latter may however lack in young wood initials or in juniper thickets. In the pasture-woodlands, such zonation complexes have developed as a consequence of anthropo-zoogenic influence; this is meanwhile generally acknowledged (cf. among others Tüxen 1967; Jakucs 1970; Burrichter *et al.* 1980). The shrubs are composed - depending on the substrate and landscape type - of different bushes, which are avoided by the grazing cattle, as they are thorny, prickly, or aromatic.

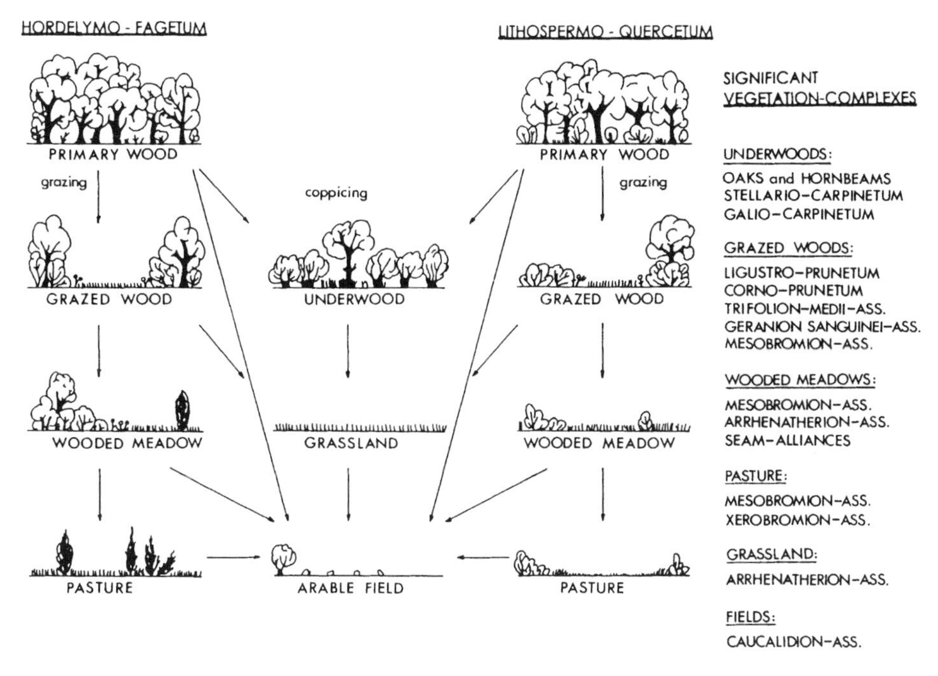

Figure 15. Degradation scheme and anthropo-zooogenic diversity of woodlands on limestone, due to extensive management and land-use.

These species are, along with the damage caused by the browsing cattle, responsible for the dynamic processes of degradation and regeneration, and explain the physiognomy of the pasture-woodlands. They also account for the changing aspects in different time intervals, mainly caused by dislocation, new formation, and destruction of the wood clusters by the grazing cattle. The pasture-woodland is not a static, but highly dynamic landscape. The grazing intensities and modalities, as well as the size of the cattle herds are therefore essential for its durable preservation, care and protection.

The traditional cattle droving, practised since the Middle Ages by migrating shepherds in the form of transhumances, has virtually ceased to exist. It certainly had an important function for the conservation of numerous pathways in the form of pasture lands with characteristic mesoxerophytic and rough meadows. An extensive network of pathways and routes, on which the cattle was driven from place to place, led across harvested fields and broad stretches of green land, along hedges and forest plots. The migratory sheep breeding was in many regions the most common and important form (see Fig. 19). In its course, the sheep transported a multitude of seeds of characteristic dry meadow elements; the seeds clung on burs to their skin or were caught between their hooves, and subsequently dislocated. The importance of this for the floral composition of the grassland communities and for the long-distance spreading of corresponding species should not be underestimated.

Figure 16. Oak-birch forest landscape as pasture-woodland, showing, as a consequence of the influence of grazing, accumulations of *Ilex aquifolium* and *Juniperus communis*.

Vegetation and physiognomy of pasture-woodlands are characterized by both the natural site conditions and the respective grazing intensities and modalities. Since these factors may change within certain regions or time periods, there is no uniform pasture-woodland type, however, comparable features can be detected, as a consequence of the effects of grazing and pasture selection. This has been repeatedly documented and demonstrated at many examples in basic works on the phytosociological registration and description of north-west German pasture-woodlands (Burrichter *et al.* 1980; Pott [*e.g.* 1982, 1985]; Pott & Hüppe 1991).

Figure 17. Shrubs rich in *Prunus spinosa*, as an example for zoogenic formations in a pasture-woodland complex of a *Carpinion*-valley plain landscape in the "Borkener Paradies" (1980).

Figure 18. Former, little utilized pasture-woodland, the tree stratum of which contains old beeches (taken from Pott & Hüppe 1991).

Figure 19. Heathland near Haltern in Westphalia, comprising *Calluna* heath and juniper instead of the oak, birch, or beech-oak forests, which originally grew there on acid soils (photo archives of the Westphalian Museum for Biology, Münster).

References

Aaby, B. 1983. Forest development, soil genesis and human activity illustrated by pollen and hypha analysis of two neighbouring podzols in Draved Forest, Denmark. Danm. Geol. Untersog. II 114: 1-114.

Arndt, E. 1989. Beiträge zur Insektenfauna der DDR: Gattung *Carabus* Linné (Coleoptera: Carabidae). Beitr. Ent. 39: 63-103.

Aßmann, T. 1990a. Polyallelic genes in the carabid beetle *Carabus punctatoauratus* Germar from the Pyrenées. pp. 319-324. In: N. E. Stork (ed); The role of ground beetles in ecological and environmental studies. Intercept, Hampshire.

Aßmann, T. 1990b. Genetische Differenzierung bei den Laufkäfern *Carabus punctatoauratus* Germar und *Carabus auronitens* Fabricius (Col. Carabidae). Verh. Westd. Entom. Tag. 1989: 5-6.

Aßmann, T. 1991. Die ripikole Carabidenfauna der Ems zwischen Lingen und dem Dollart. Osnabrücker naturwiss. Mitt. 17: 95-112.

Aßmann, T. 1994. Epigäische Coleopteren als Indikatoren für historisch alte Wälder der Nordwestdeutschen Tiefebene. Ber. NNA 3: 142-151

Aßmann, T., O. Nolte &. Reuter, H. 1994. Postglacial colonization of middle Europe by *Carabus auronitens* F. as revealed by population genetics (Coleoptera, Carabidae). pp. 3-9. In: Desender, K., Dufrêne, M., Loreau, M., Luff, M. L. & Maelfait, J.-P. (eds), Carabid beetles, ecology and evolution. Kluwer Academic Press, Dordrecht, Boston, London:

Aßmann, T. & Kratochwil, A. 1995. Biozönotische Untersuchungen in Hudelandschaften Nordwestdeutschlands - Grundlagen und erste Ergebnisse. Osnabrücker Naturwiss. Nachr. 20/21: 275-337.

Bathke, M. 1994. Tagfalter-Biozönosen im Hudelandschaftsmosaik “Borkener Paradies” (Emsland). Diploma thesis, Univ. Osnabrück.

Brunet, J. 1992. Impact of grazing on the field layer vegetation in a mixed oak forest in south Sweden. Svensk Bot. Tidskr. 86: 347-353.

Burrichter, E. 1984. Baumformen als Relikte ehemaliger Extensivwirtschaft in Nordwestdeutschland. Drosera 84: 1-18

Burrichter, E., Pott, R., Raus, T. & Wittig, R. 1980. Die Hudelandschaft “Borkener Paradies” im Emstal bei Meppen. - Abh. Landesmus. f. Naturk. 42: 1-69, Münster.

Burschel, P. & Huss, J. 1987. Grundriß des Waldbaus. Ein Leitfaden für Studium und Praxis. Pareys Studientexte 49. Parey Verlag, Hamburg und Berlin.

Dinter, W. 1991. Die floristische Sonderstellung alter Wälder im Tiefland Nordrhein-Westfalens: das Beispiel des Hiesfelder Waldes. Geobot. Kolloq. 7: 83-84.

Elerie, J.N.H., Jager, S.W. & Spek, Th. 1993. Landschaps-geschieden is van de Strubben/Kuiphorstbos. Archaeologische en historisch-ecologische Studies van een Naturgebied op de Hondsrng. Van Dijk Foosthuis Regio-Projekt-Groningen.

Ellenberg, H. 1986. Vegetation Mitteleuropas mit den Alpen in ökologischer Sicht. Ulmer Verlag, Stuttgart.

Emanuelson, U. 1988. The relationship of different agricultural systems to the forest and woodlands of Europe. pp. 169-178. In: Salbitano, F. (ed), Human influence on forest ecosystems development in Europe. Edf Fern-Cnr, Pitagora Editrice, Bologna.

Emanuelson, U. Bergendorff, C., Carlsson, B., Lewan, N. & Nordell, O., 1985. Det skanska kulturlandskapet. Signum, Lund.

Falinski, J. B. 1986. Vegetation dynamics in temperate lowland primeval forests. Geobotany 8.

Feldmann, R. 1981. Die Amphibien und Reptilien Westfalens. Abh. Landesmus. Naturkde. Münster 43: 1-161.

Härdtle, W. 1995. Vegetation und Standort der Laubwaldgesellschaften (*Querco-Fagetea*) im nördlichen Schleswig-Holstein. Mitt. d. Arbeitsgem. Geobotanik in Schleswig-Holstein und Hamburg 48: 1-441.

Hermy, M. 1989. Former land use and its effects on the composition and diversity of woodland communities in the western part of Belgium. Stud. plant ecol. 18: 104. Uppsala.

Hermy, M. & Stieperaere, H. 1981. An indirect gradient analysis of the ecological relationship between ancient and recent riverine woodlands to the south of Breges (Flanders, Belgium). Vegetatio 44: 43-49.

Hesmer, H. & Schröder, F.G. 1963. Waldzusammensetzung und Waldbehandlung im Niedersächsischen Tiefland westlich der Weser und in der Münsterschen Bucht bis zum Ende des 18. Jahrhunderts. Decheniana. Beih. 11: 1- 304.

Hockmann, P., Menke, K., Schlomberg, P. & Weber, F. 1992. Untersuchungen zum individuellen Verhalten (Orientierung und Aktivität) des Laufkäfers *Carabus nemoralis* im natürlichen Habitat. Abh. Westf. Museum Naturkde. 54: 65-98.

Inghe, O. & Tamm, C. O. 1985. Survival and flowering of perennial herbs. IV. The behavior of *Hepatica nobilis* and *Sanicula europaea* on permanent plots during 1943-1981. Oikos 45: 400-420.

Isenberg, E. 1979. Pollenanalytische Untersuchungen zur Vegetations- und Siedlungsgeschichte in der Grafschaft Bentheim. Abhandl. Landesmus. Naturk. 41: 1-63.

Jakucs, P. 1970. Die Sproßkolonien und ihre Bedeutung in der dynamischen Vegetationsentwicklung (Polycormonsukzessionen). Acta Bot. Croat. 28: 161-170.

Koop, H. 1989. Forest Dynamics, SILVI-STAR: A comprehensive monitoring system. Springer-Verlag, Heidelberg.

Koop, H. 1991. Untersuchungen der Waldstruktur und der Vegetation in den Kernflächen niederländischer Naturwaldreservate. Schriftenr. Veg.kde.: 21: 67-76.

Korpel, S. 1995. Die Urwälder der Westkarpaten. G. Fischer-Verlag, Stuttgart-Jena, New York.

Kratochwil, A. 1983. Zur Phänologie von Pflanzen und blütenbesuchenden Insekten (Hymenoptera, Lepidoptera, Diptera, Coleoptera) eines versaumten Halbtrockenrasens im Kaiserstuhl - Ein Beitrag zur Erhaltung brachliegender Wiesen als Lizenzbiotope gefährdeter Tierarten. Beih. Veröff. Naturschutz Landschaftspflege Bad.-Württ. 34: 57-108.

Kratochwil, A. 1984. Pflanzengesellschaften und Blütenbesuchergemeinschaften: biozönologische Untersuchungen in einem nicht mehr bewirtschafteten Halbtrockenrasen (*Mesobrometum*) im Kaiserstuhl (Südwestdeutschland). Phytocoenologia 11: 455-669.

Kratochwil, A. 1988. Tagung des Arbeitskreises "Biozönologie" in der Gesellschaft für Ökologie am 14. und 15. Mai 1988 in Freiburg i. Br. - Einführung; Verlauf und Resümee. Mitt. Bad. Landesv. Naturkde. u. Natursch. N.F. 14: 537-546.

Kratochwil, A. 1993. Forschungsbericht 1992/93, Fachgebiet Ökologie am Fachbereich Biologie/Chemie der Univ. Osnabrück.

Kratochwil, A. & Aßmann, T. 1996. Biozönotische Konnexe im Vegetationsmosaik nordwestdeutscher Hudelandschaften. Ber. d. Reinh.-Tüxen-Ges. 8: 237-282

Leibundgut, H. 1993. Europäische Urwälder. Haupt-Verlag, Bern u. Stuttgart.

Mantel, K. 1990. Wald und Forst in der Geschichte. Alfeld-Hannover.

Mayer, H., Zukrigl, W. Schrempf, W. & Schlager, G. 1987. Urwaldreste, Naturwaldreste und schützenswerte Naturwälder in Österreich. Inst. f. Waldbau und Bodenkultur, Wien.

Niemann, A. 1994. Bestandsgrößen, Habitatwahl und Habitatnutzung heimischer Meisen (Paridae und Aegithalidae) in emsländischen Hudelandschaften Wacholderhain "Wachendorf", Meppener Kuhweide. Diploma thesis, Univ. Osnabrück.

O'Connell, M. 1986. Pollenanalytische Untersuchungen zur Vegetations- und Siedlungsgeschichte aus dem Lengener Moor, Friesland. Probleme der Küstenforschung 16: 171-193.

Peterken, G.F. & Game, M. 1981. Historical factors affecting the distribution of *Mercurialis perennis* in central Lincolnshire. J. Ecol. 69: 781-796.

Peterken, G.F. & Game, M., 1984. Historical factors affecting the number and distribution of vascular plant species in the woodlands of central Lincolnshire. J. Ecol. 72: 155-182.

Peterken, G.F. 1977. Habitat conservation priorities in British and European woodlands. Biol.Cons. 11: 223-236.

Peterken, G.F. 1981. Woodland conservation and management. Chapman & Hall., London.

Peterken, G. F. 1988. Use of history of individual woods in modern nature conservation. pp. 201-214. In: Salbitano, F. (ed), Human influence on forest ecosystems development in Europe. Edf Fern-Cnr, Pitagora Editrice, Bologna.

Plachter, H. 1992. Der Beitrag von Arten- und Biotopschutz-Programmen zu einem zeitgemäßen Naturschutz. Schriftenr. Bayer. Landesamt f. Umweltschutz, Heft 100: 15-22.

Pott, R. 1981. Der Einfluß der Niederwaldwirtschaft auf die Physiognomie und die floristisch-soziologische Struktur von Kalkbuchenwäldern. Tuexenia 1: 233-242.

Pott, R. 1985. Vegetationsgeschichtliche und pflanzensoziologische Untersuchungen zur Niederwaldwirtschaft in Westfalen. Abh. Westf. Mus. f. Naturk. 47: 1-75.

Pott, R. 1988a. Entstehung von Vegetationstypen und Pflanzengesellschaften unter dem Einfluß des Menschen. Düsseld. Geobot. Kolloq. 5: 27-54.

Pott, R. 1988b. Impact of human influences by extensive woodland management and former land-use in North-Western Europe. pp. 263-278. In: Salbitano, F. (ed), Human influence on forest ecosystems development in Europe. Esf Fern-Cnr, Pitagora Editrice, Bologna.

Pott, R. 1993. Farbatlas Waldlandschaften. Ausgewählte Waldtypen und Waldgesellschaften unter dem Einfluß des Menschen. Ulmer Verlag, Stuttgart.

Pott, R. 1994. Naturnahe Altwälder und deren Schutzwürdigkeit. Ber. NNA 7: 115-133.
Pott, R. 1995. Die Pflanzengesellschaften Deutschlands. Ulmer Verlag, Stuttgart.
Pott, R. 1996. Biotoptypen. Schützenswerte Lebensräume Deutschlands und angrenzender Regionen. Ulmer Verlag, Stuttgart.
Pott, R. & Burrichter, E. 1983. Der Bentheimer Wald - Geschichte, Physiognomie und Vegetation eines ehemaligen Hude- und Schneitelwaldes. Forstwiss. Centralblatt 102: 350-361.
Pott, R. & Hüppe, J. 1991. Die Hudelandschaften Nordwestdeutschlands. Abh. Westf. Mus. f. Naturk. 53: 1-314.
Prusa, E. 1985. Die böhmischen und mährischen Urwälder. Vegetace CSSR, 15. Verlag der Tschechoslowakischen Akademie der Wissenschaften. Prague.
Rackham, O. 1976. Trees and woodland in the British landscape. London.
Rackham, O. 1980. Ancient woodland. Its history, vegetation and uses in England. Edward Arnold, London.
Salbitano, F. (ed) 1988. Human influence on forest ecosystems development in Europe. ESF. Pitagora Editrice, Bologna.
Schuler, A. 1988. Forest area and forest utilization in the Swiss pre-alpine region. pp. 121-128. In: Salbitano, F. (ed), Human influence on forest ecosystems development in Europe. Esf Fern-Cnr, Pitagora Editrice, Bologna.
Teller, A. 1990. Directory of European research groups active in forest ecosystem research (Fern). European Science Foundation, Strasbourg.
Tüxen, R. 1967. Die Lüneburger Heide. Werden und Vergehen einer Landschaft. Rotenburger Schr. 26: 1-52.
Watkins, C. 1988. The idea of ancient woodland in Britain from 1800. pp. 237-246. In: Salbitano, F. (ed). Human influence on forest ecosystems development in Europe. Esf Fern-Cnr, Pitagora Editrice, Bologna.
Wittig, R. 1991. Schutzwürdige Waldtypen in Nordrhein-Westfalen. Geobot. Kolloq. 7: 3-15.
Zacharias, D. 1993. Zum Pflanzenartenschutz in Wäldern Niedersachsens. Ber. NNA 4: 21-29.
Zacharias, D. & Brandes, D. 1989. Floristical data analysis of 44 isolated woods in northwestern Germany. Stud. Plant. Ecol. 18: 278-280.
Zacharias, D. & Brandes, D. 1990. Species area-relationships and frequency - Floristical data analysis of 44 isolated woods in northwestern Germany. Vegetatio 88: 21-29.
Zukrigl, K., Eckhardt, G. & Nather, J. 1963. Standortskundliche und waldbauliche Untersuchungen in Urwaldresten der niederösterreichischen Kalkalpen. Mitt. Forstl. Bundesversuchsanstalt Mariabrunn : 1-244.

PLANT DIVERSITY IN SUCCESSIONAL GRASSLANDS: HOW IS IT MODIFIED BY FOLIAR INSECT HERBIVORY?

VALERIE K. BROWN[1] & ALAN C. GANGE[2]

Imperial College at Silwood Park, Ascot, Berks. SL5 7PY, UK
Current Addresses: [1] *CABI Biosience: Environment UK Centre (Ascot), Silwood Park, Ascot, Berks, UK and* [2] *School of Biological Sciences, Royal Holloway, Egham, Surrey TW20 0EX, UK*

Key words: grassland, insect herbivory, plant diversity, succession

Introduction

Plant species diversity varies during ecological succession and this change is invoked by a wide range of abiotic and biotic factors. The more commonly cited biotic factors include the role of the seed bank, (the initial floristic composition of Egler (1954), changing life-history characteristics (Brown & Southwood 1987), competitive ability of the plant species (Wilson & Shure 1993) and the role of associated organisms, namely pathogenic or mutualistic fungi (Gange *et al.* 1990) or vertebrate or invertebrate herbivores (Brown & Gange 1992; Gibson *et al.*1987). In this chapter, we focus on the direct effects of herbivory, whilst accepting that it is seldom possible to disentangle these from indirect effects. The latter are commonly mediated via differential herbivory on plant species or the differential responses of plant species to herbivory. As the plant species change during succession, the relative balance of direct and indirect effects of herbivory are likely to change. However, this involves a major assumption, namely that the herbivores remain the same. While this may be true for vertebrate herbivores, especially when used in management, it is far from true for the smaller invertebrate herbivores, which themselves undergo successional dynamics (Brown & Southwood 1987).

The role of various types of herbivores in directing plant succession is reviewed by Luken (1990). Although the effects of herbivory on plant community structure are considered, emphasis is on the relevance of the activities of vertebrate herbivores in managing succession. Thus, invertebrate herbivores are not even mentioned! Indeed, the smaller size of invertebrates and, in general, their more sedentary behaviour has resulted in claims that they invoke only negligible or subtle changes in vegetation (Crawley 1989). However, many invertebrates occur at large population densities and display selectivity in the way they exploit plants as a food resource and in the species of plant eaten. Thus, invertebrates have the ability to bring about changes in the growth, performance and fitness of individual species and in so doing modify the competitive balance between plant species. In this way, plant diversity and community dynamics

A. Kratochwil (ed.), Biodiversity in ecosystems, 133-146

can be changed, (*e.g.* Brown 1990; Brown & Gange 1992; McBrien *et al.* 1983; McEvoy *et al.* 1993). Indeed, the invertebrates involved may be considered as "keystone consumers" in successional dynamics (Davidson 1993).

Davidson (1993) has considered the evidence for herbivores modifying successional pathways and explained these in terms of resource availability models of plant defence investment. From these models, it has been predicted that herbivores should feed more on pioneer than later successional species, and therefore accelerate the demise of the pioneers, serving to hasten successional change. This has indeed been recorded in some cases (*e.g.* Bryant & Chapin 1986 and McBrien *et al.* 1983). However, there appear to be as many exceptions recorded, namely when herbivores feed preferentially on later successional species and therefore retard the rate of succession (*e.g.* Brown & Gange 1992; Root 1996). The explanation for these effects appears to rest more with the attributes of the plants preferentially eaten than those of the herbivores (Davidson 1993). It is, therefore, the position in the successional sequence of the most palatable species that governs the impact of herbivory on succession. However, the feeding strategy, host plant preference and population dynamics of the insects must be of major significance. Unfortunately, this part of the equation is often ignored. Furthermore, in Davidson's (1993) model of succession, the herbivore effect at the annual plant/perennial forb transition stage is omitted, through lack of knowledge. It is our aim in this chapter to provide data to fill this gap.

In an experiment at Silwood Park, Berkshire, UK, we used a series of sites of known and different successional age, but which were established in the same way. The experiment involved the reduction of insect herbivores by the judicious use of insecticides and comparison with vegetation subjected to natural levels of insect herbivory. Since the composition of the vegetation and the insect herbivores in each site were known in detail (Brown & Southwood 1987), we can use this system to (i) assess the effect of foliar insect herbivory on plant species diversity, (ii) show how the nature and magnitude of the effects vary with the successional age of the community and of the insect herbivore communities, (iii) explore the mechanism underpinning the effects and finally (iv) test the prediction that it is the stage in succession at which the most palatable species occurs that governs whether succession is retarded or accelerated.

The experimental system

Field sites

Within a large area of ex-arable land, on sandy acidic soil, at Silwood Park, Berkshire, UK, a series of sites approximately 30 m x 20 m have been established in a similar way over a period of fifteen years. Each site was treated with weedkiller in autumn to destroy perennial weeds, shallow ploughed in winter, harrowed and hand-raked in early spring and the vegetation left to develop naturally. The experiment described here employs sites created in 1986, 1983 and 1979, in addition to a site of mature grassland (adjacent to the other sites) which had been grazed by rabbits before the start of the experiment, when all sites were fenced. This selection of sites provides representative examples of four main successional communities. The site created in 1986 represents ruderal community, dominated by annual and short-lived perennial forbs, that created in 1983 an early successional community in which perennial forbs (non graminaceous species) and perennial grasses start to colonise, that created in 1979 an early-mid successional community with a mixture of perennial forbs and grasses, while the pasture represents a mid successional grassland in which perennial grasses dominate. Within each site, twenty 3 m x 3 m plots were defined. Each plot was separated from its neighbour by at least 2 m and systematically allocated to a treatment: 10 plots ('insecticide-treated') were sprayed at three weekly intervals from May - October with 750 ml Dimethoate-40, a contact and systemic insecticide, at the recommended agricultural rate of 0.336 kg a.i.ha, the other 10 with 750 ml water ('control'). Detailed tests for direct effects of the chemical on the vegetation, using methodology described in Brown *et al.* (1987) and Gange *et al.* (1992) revealed no significant effects. These treatments began immediately after site preparation in the ruderal site and in spring 1985 in the other three sites.

Sampling

The vegetation developing on the ruderal site (where changes were most rapid) was sampled at fortnightly intervals from April to October in the first season (1986) and at monthly intervals in subsequent seasons and all sites. Five linear frames, each containing ten 3 mm diameter point quadrat pins, were randomly placed in each plot. The number of touches of all living plant material was recorded in 2 cm (below 10 cm) or 5 cm (10 cm and above) height intervals. These data provided information on plant species richness (number of plant species per plot), and cover abundance (total number of touches of all species at all height intervals) and the cover abundance of plant species belonging to different life-history groupings (*e.g.* annual forbs, perennial forbs, perennial grasses). Data were analysed according to dates with a split-plot Analysis of Variance (ANOVA), with the insecticide treatment as the main effect. All data satisfied the assumptions of normality and homogeneity of variances.

Results

The insect fauna

The insects associated with the differently aged sites have been studied in detail, with particular reference to the herbivorous fauna and to the adaptive life-history traits of the species (Brown & Southwood 1987; Edward-Jones 1991). Foliar herbivore species richness and density is typically low in the ruderal stages of succession, though some species may be locally abundant (Stinson 1983; Edwards-Jones & Brown 1993). Species diversity reaches its highest levels in early succession and then declines gradually (Fig. 1a). However, the species composition, as displayed by measures of beta diversity, varies considerably throughout succession (Brown & Southwood 1987). In addition, and probably more importantly as far as effects on plants are concerned, the feeding strategy of the insects also varies (Brown 1986, 1990b). Species associated with the ruderal plant community are mainly generalist forb feeders, whilst those establishing in later succession are more specialist grass feeders, including leaf miners and gall formers. From these dynamics of the insect community, we predict greater effect on annual forbs in ruderal succession, switching to more substantial impacts on perennial grasses in early-mid succession. Clearly, the magnitude of these effects will be exacerbated by the relative abundance of the plant life-history groupings at various stages of succession. We now look to test these predictions.

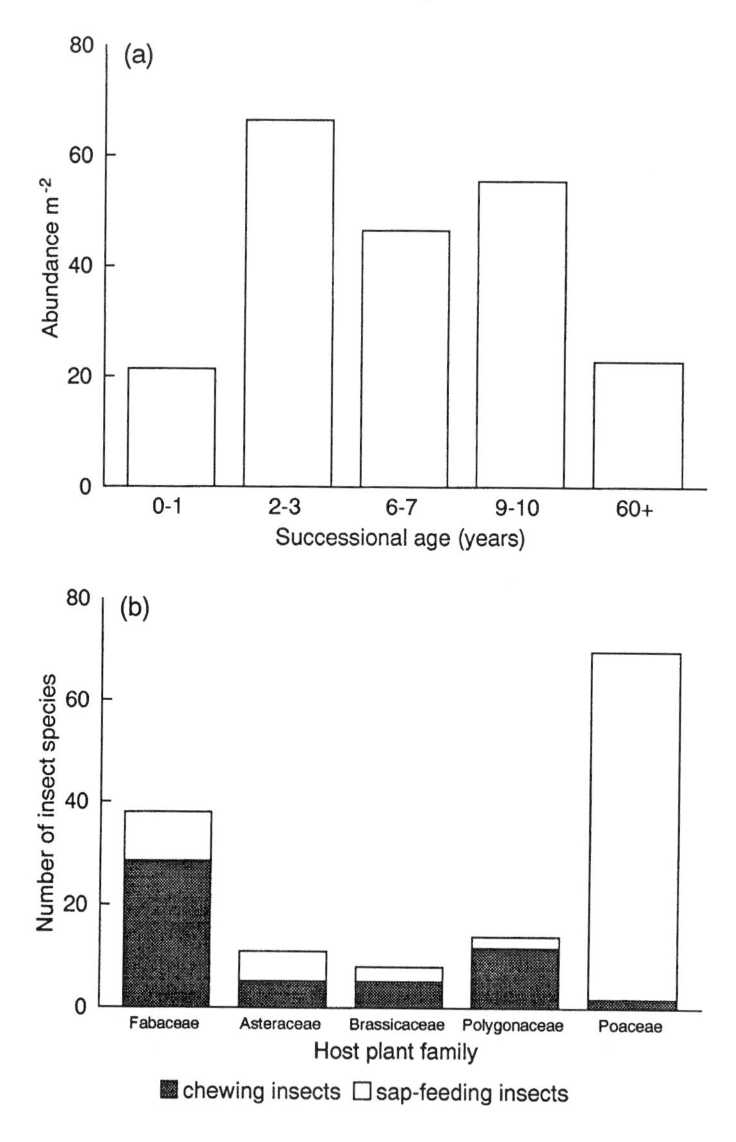

Figure 1. a) Foliar-feeding insect density in relation to the successional age of the plant community; b) Number of foliar-feeding insect species on plants from different families (from Edwards-Jones 1988).

In a comparison of the foliar herbivore load on plants of different successional age (Edwards-Jones 1988; Edwards-Jones & Brown 1993), major differences were found on different plant families and species which typically have a characteristic place in succession. For example, over sixty insect species were recorded feeding on Poaceae, while the next most herbivore rich family was the Fabaceae with 35 species, with other families having approximately half this number (Fig. 1b). Furthermore, the graminaceous feeders were mainly restricted to early and mid succession. The majority of these species were Homoptera, belonging to the families Cicadellidae, Delphacidae and Cercopidae, which feed on the phloem of plant tissues which they exploit by specialised sucking mouthparts. Many species show a preference for particular grass species, commonly *Holcus lanatus* and *Agrostis capillaris* L.

Plant species diversity

Plant species diversity, typically recorded as species richness, on acidic sandy soils in southern England is typically lower than on mesic or calcareous soils (Brown *et al.* 1987; Gibson & Brown 1991). However, there is a distinct successional trend which is consistent from year to year. Typically, species richness was high in the first two years of colonisation of bare ground, when annual forb species were abundant and perennial forbs and grasses were colonising. However, this leveled off in early succession and declined in the mid successional pastures, once the highly competitive grasses became established (Southwood *et al.* 1979).

The modification in species richness brought about by the application of foliar insecticide provided a measure of the impact of these insects on this community parameter. In the ruderal site, plant species richness was significantly increased in the first year of colonisation (and of treatment) (Fig. 2). Thereafter, and in all other sites, there were generally fewer plant species in the insecticide-treated plots. However, this effect was only significant from the second year of treatment (early-mid successional site), third year (early successional site) or fourth year in the pasture, where the species richness was lowest (see Table 1). Fig. 3, showing the early successional site, was typical of the response of the older sites to insecticide treatment. The greatest differences tended to occur in the intermediate-aged sites after several years of treatment.

Vegetation cover and composition

A commonly used measure of vegetation cover, referred to here as frequency, is given by the number of pins touched by vegetation. A more sensitive measure, however, is the number of touches of vegetation on single pins. This affords a measure of cover abundance, and takes into account the differential structural complexity of individual species. With the exception of the established sites in their first years of treatment, for the 21 sample years, spanning four sites and five or six years of sampling on each site, there was a higher cover abundance in the insecticide-treated than in the control plots. In the ruderal site, the difference was highly significant from the first year of vegetation colonisation and, in the more mature sites, significant differences were consistently seen from the second year of treatment (Table 2). Fig. 4, depicting the trends in the early-mid successional site, is typical of these latter sites. There was a clear mid season peak in vegetation cover in insecticide-treated plots (Fig. 4), giving rise to highly significant date and date x treatment interaction terms in the ANOVA. Cover abundance was reduced in Year 5 due to an unusually dry summer.

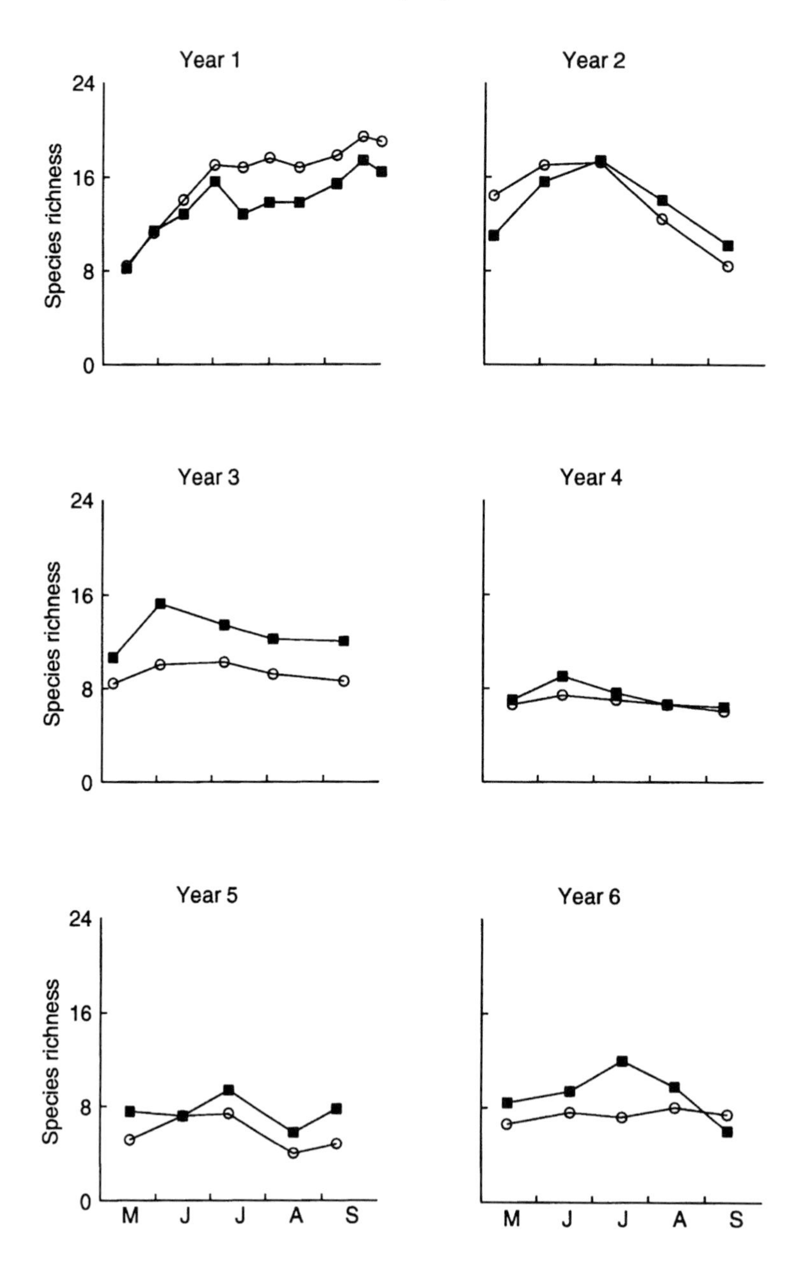

Figure 2. Plant species richness in control (■) and foliar insecticide-treated (O) plots during the first six years of colonisation of bare ground. Results from ANOVA are given in Table 1.

Table 1. Summary of ANOVA values comparing plant species richness in four sites of different successional age when treated with foliar insecticide for at least five years. Degrees of freedom for all F values; 1,16 (see text)

	Year of Treatment					
Site	1	2	3	4	5	6
Ruderal	25.5 [1]	0.1	27.8 [3]	1.8	27.3 [3]	1.12
Early	0.0	0.8	30.0 [3]	27.6 [3]	63.3 [3]	-
Early-mid	0.2	4.5 [1]	1.0	45.0 [3]	26.0 [3]	-
Mid	1.4	1.3	1.2	7.5 [1]	7.0 [1]	-

[1] $p < 0.05$, [2] $p < 0.01$, [3] $p < 0.001$.

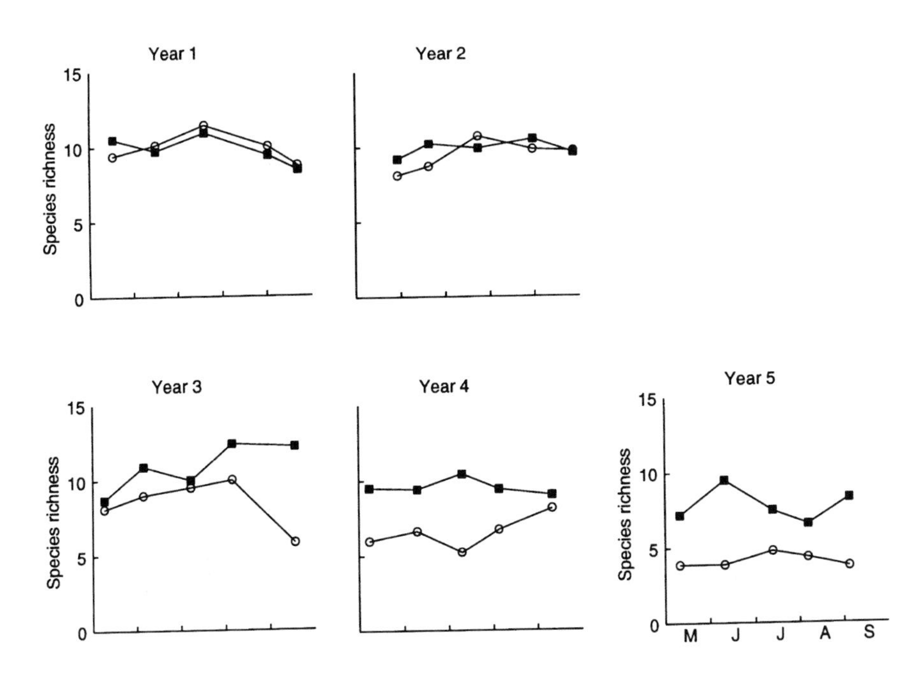

Figure 3. Plant species richness in control (■) and foliar insecticide-treated plots (O) in the early successional site for five consecutive years, following the commencement of experimental treatment. Results from ANOVA are given in Table 1. Note 'Year 1' represents the third growing season of the site.

Table 2. Summary of ANOVA values comparing cover abundance of vegetation in four sites of difference successional age when treated with insecticide for at least five years. Degrees of freedom for all F values: 1,16 (see text)

Site	Year of Treatment					
	1	2	3	4	5	6
Ruderal	105.5 [3]	117.7 [3]	68.5 [3]	29.8 [3]	66.4 [3]	10.1 [1]
Early	0.0	15.5 [2]	61.8 [3]	43.4 [3]	28.1 [3]	-
Early-mid	1.4	99.7 [3]	117.0 [3]	118.7 [3]	70.2 [3]	-
Mid	0.1	13.1 [3]	82.2 [3]	46.9 [3]	27.3 [3]	-

[1] $p < 0.05$, [2] $p < 0.01$, [3] $p < 0.001$.

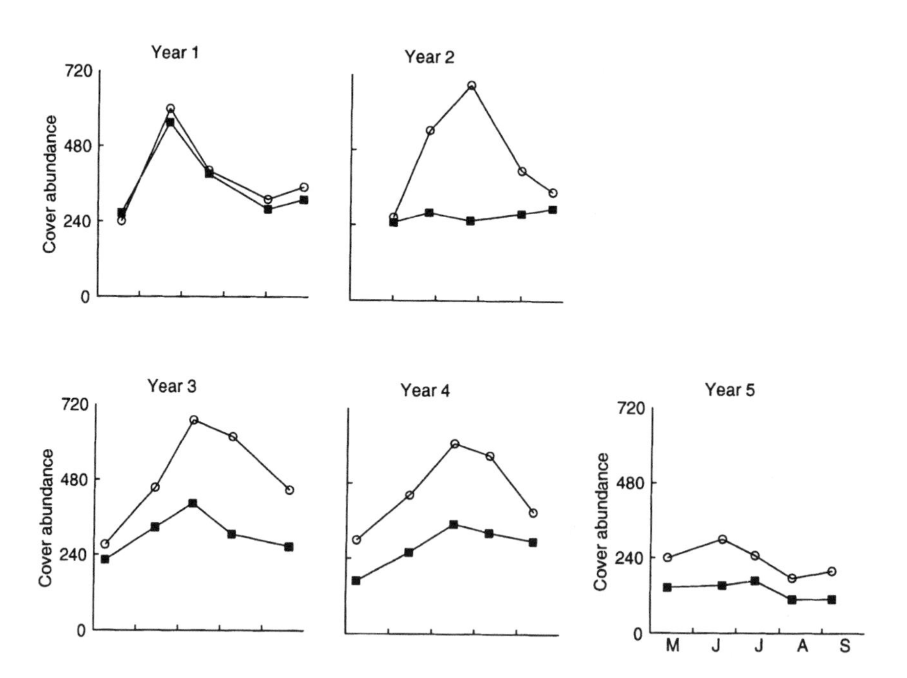

Figure 4. Cover abundance of vegetation in control (■) and foliar insecticide-treated (O) plots in the early-mid successional site for five consecutive years, following the commencement of experimental treatment. Results from ANOVA are given in Table 2. Note 'Year 5' represents the seventh growing season of the site.

Plant species were grouped according to grass or forb and annual or perennial. This was possible because there were very few monocarpic perennial species or other plant types, such as sedges or rushes. Indeed, there were too few annual grasses to analyse, with *Poa annua* being the only species represented by more than a few individuals in the youngest site.

Annual forb species only contributed substantially to the vegetation during the first two years of colonisation of bare ground. They were significantly enhanced by the application of insecticide in the first year of colonisation ($F_{1,16} = 95.8$, $p < 0.001$). This result is most likely explained by the higher seedling survival in the insecticide-treated plots (Brown 1982; Brown & Gange 1989). However, a few annual forbs did persist (or recolonised) in the older sites and these were significantly more abundant in control plots.

Perennial forbs started to colonise the ruderal site in the autumn of the first year of colonisation, but on average only constituted a small proportion of the vegetation cover at any stage in the succession, and never more than 30 %. In general, the cover abundance of the group was depressed by insecticide treatment, although in the second year of treatment in the early and early-mid successional site they were enhanced (early $F_{1,16} = 5.4$, $p < 0.05$; early-mid $F_{1,16} = 4.4$, $p < 0.05$). The enhancement suddenly faded by mid season of the third year of treatment, such that their cover was below that of control plots by the end of the season. This resulted in very highly significant treatment x data interactions in both sites ($p < 0.001$). In the mid successional site, the enhancement was found one year later in the insecticide treated plots and was brought about by an increase in flowering of *Galium saxatile*.

The perennial grasses were the dominant constituent of the vegetation in all successional sites and this group showed the strongest response to insecticide treatment. Like the perennial forbs, they became established in the second year of colonisation in the ruderal site and for the remaining five years of the experiment were significantly enhanced (all $p < 0.001$) by the application of insecticide. In the older sites, the response to insecticide application was consistent, viz. no response in the first year of treatment, followed by consistently significant differences in the early and early-mid successional sites. In the mid successional site, the same trend was seen, although the difference developed more slowly. Fig. 5 shows the trend for the early successional site and is representative of the other sites.

In all except the early years of colonisation of bare ground, it was the balance between the perennial grasses and forbs which was most affected by the application of insecticide. This balance is best displayed by the grass/forb ratio, based on cover abundance estimates for the perennial grasses in relation to all forbs. A ratio of unity therefore indicates equal representation of both groups. Fig. 6 shows the ratio for the early successional site. It can be seen that the ratio varied little from unity in the first two years of treatment, but increased in insecticide treated plots at the end of the third year. Thereafter, the ratio was significantly elevated in treated plots (Year 4: $F_{1,16} = 12.7$, $p < 0.01$; Year 5: $F_{1,16} = 9.2$, $p < 0.01$).

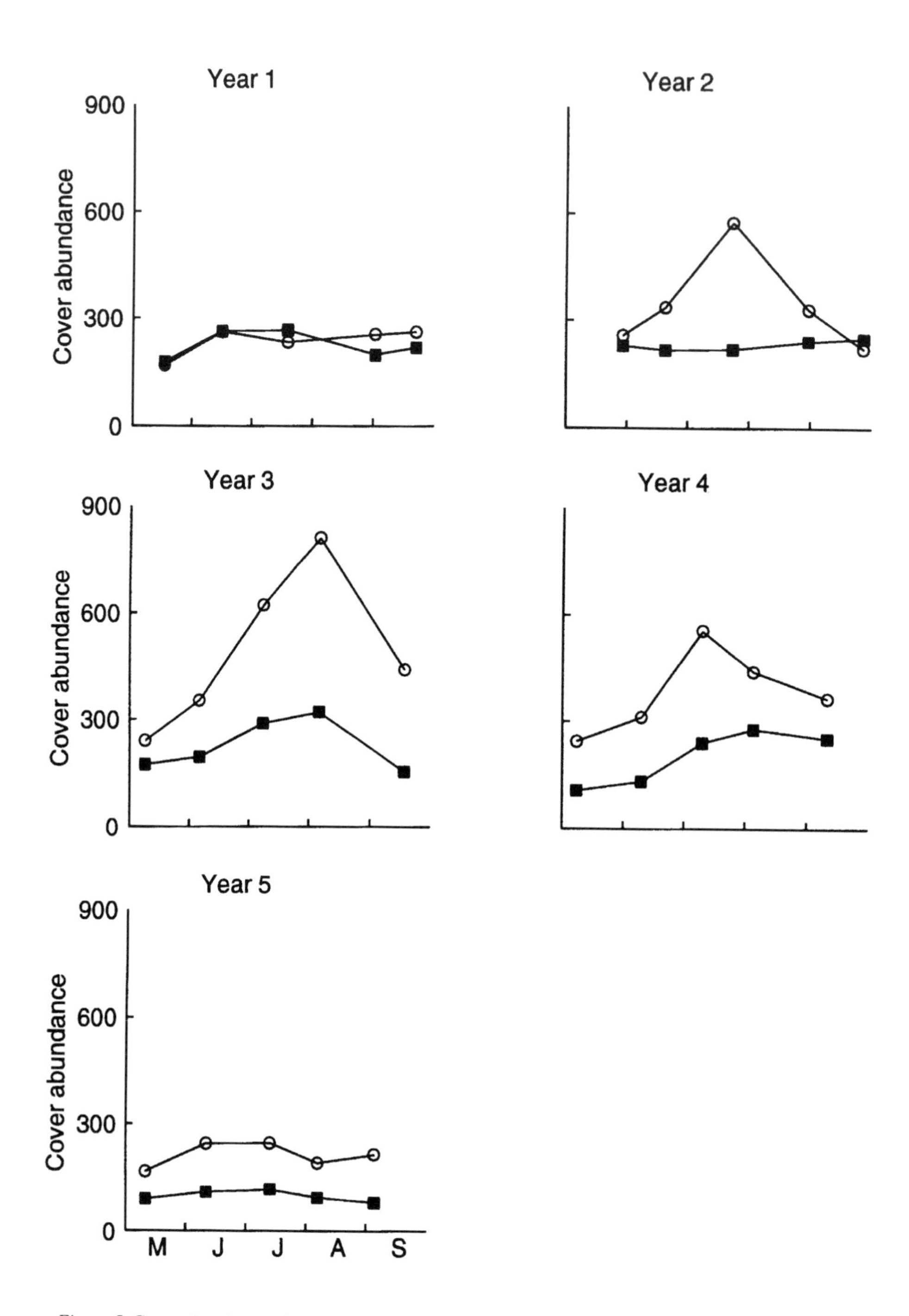

Figure 5. Cover abundance of perennial grasses in control (■) and foliar insecticide-treated (O) plots in the early successional site for five consecutive years, following the commencement of treatment. Note 'Year 1' represents the third growing season of the site.

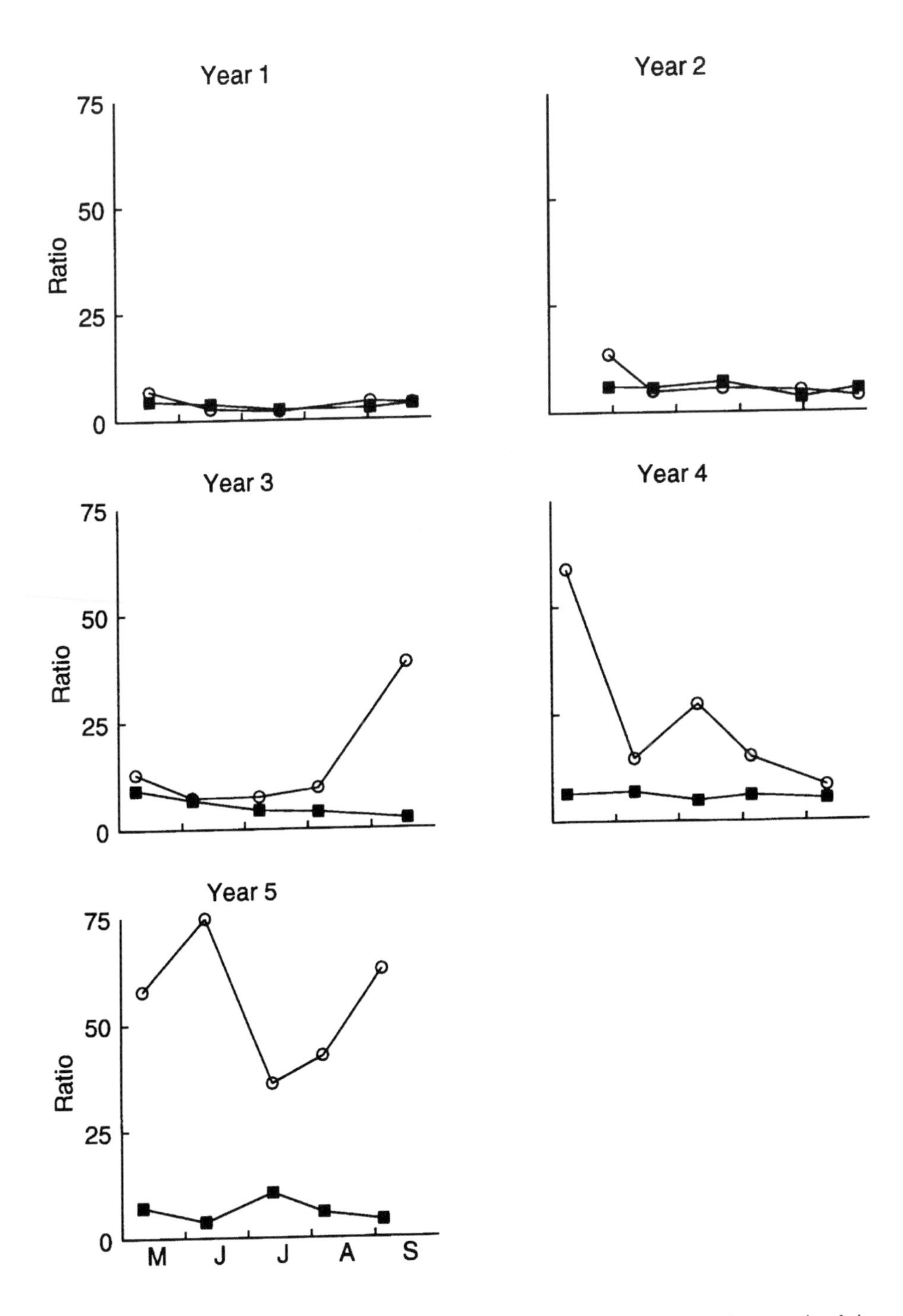

Figure 6. Grass/forb ratio in control (■) and foliar insecticide-treated (O) plots in the early successional site for five consecutive years, following the commencement of treatment. Note 'Year 1' represents the third growing season of the site.

The mechanism of changing plant diversity

The results described show a clear response of plant communities of different successional age to a reduction in above-ground herbivory by insects. However, the results display interesting differences in the response of vegetation at different stages in succession. These differences are explicable by the nature of the insect herbivore community at particular stages in succession (Fig 1). In the ruderal stages, the insects were mainly generalist herbivores feeding on a wide range of forb species. Consequently, it was the annual forbs which displayed the greatest response when herbivore load was reduced by insecticide application. As the plant community developed, perennial grasses became dominant and were colonised by a large number of specialist feeders which suck phloem sap. These cause no obvious damage to the plant, but its performance and competitive ability are reduced (see Gange 1990). Because grasses are generally superior competitors to forbs, any mechanism which reduces their competitive dominance will be advantageous to the persistence of the, mainly, perennial forbs. Thus, after the ruderal phase, foliar herbivory maintains the forb diversity of developing grasslands, and as such their overall species diversity.

In successional terms, the activities of above-ground insect herbivores serve to reduce the dominance of the perennial grasses and thereby retard the rate of succession to the plagioclimax community. This agrees well with Davidson's (1993) model of herbivory in succession, namely that, when the later successional species are preferentially eaten, the rate of succession is reduced. However, Davidson states that the herbivore effect on succession from annual to perennial forbs is undocumented. In the ruderal site, we found that annual forbs persisted longer in the absence of herbivores. These data therefore provide the first evidence that insect herbivores accelerate the changes from ruderal to early successional plant communities. Clearly, the nature of the prevailing insect community (Fig. 1) determines the rate and direction of successional change.

There is always a danger that the results of experimental manipulations in the field may be driven by site specific characteristics. However, this seems unlikely for the trends described here, as replication of similar experiments on different soil types in the U.K. (Gibson *et al.* 1989) and in different countries (Hendrix *et al.* 1987; Brown 1994) have produced very similar results, even though there are differences in the composition of the successional vegetation. Whereas, it is well known that grazing by larger vertebrate herbivores can maintain plant diversity (Watt 1981; Huntly 1991), we are unaware of comparable extensive studies on invertebrates, especially insects, that have displayed equally important effects.

Acknowledgements

We are grateful to the Natural Environment Research Council in the UK who funded much of the work described in this chapter. We are also appreciative of the many field assistants and students who assisted, and particularly to Gareth Edwards-Jones whose data contribute to Fig. 1.

References

Brown, V.K. 1982. The phytophagous insect community and its impact on early successional habitats. pp. 205-213. In: Visser, J.H. & Minks, A.K. (eds), Proceedings of the 5th International Symposium on Insect-Plant Relations, Wageningen 1982. Pudoc, Wageningen.

Brown, V.K. 1986. Life cycle strategies and plant succession. pp. 105-124. In: Taylor, F. & Karban, R. (eds), The Evolution of Insect Life Cycles . Springer-Verlag, New York.

Brown, V.K. 1990a. Insect herbivory and its effect on plant succession. pp. 275-287. In: Burdon, J.J. & Leather, S.R. (eds), Pests, Pathogens and Plant Communities. Blackwell Scientific Publications, Oxford.

Brown, V.K. 1990b. Insect herbivores, herbivory and plant succession. pp. 183-196. In: Gilbert, F. (ed), Insect Life Cycles: Genetics, Evolution and Co-ordination. Springer-Verlag, New York.

Brown, V.K. 1994. Herbivory: a structuring force in plant communities. pp. 299-308. In: Leather, S.R., Watt, A.D., Mills, N.J. & Walters, K.F.A. (eds), Individuals, Populations & Patterns in Ecology. Intercept Ltd, Hampshire.

Brown, V.K. & Gange, A.C. 1989. Herbivory by soil-dwelling insects depresses plant species richness. Functional Ecology 3: 667-671.

Brown, V.K. & Gange, # A. C. 1992. Secondary plant succession: How is it modified by insect herbivory? Vegetatio 101: 3-13.

Brown, V.K., Leijn, M. & Stinson, C.S.A. 1987. The experimental manipulation of insect herbivore load by the use of an insecticide (malathion): the effect of application on plant growth. Oecologia 72: 377-381.

Brown, V.K. & Southwood, T.R.E. 1987. Secondary succession: patterns and strategies. pp. 315-337. In: Gray, A.J., Crawley, M.J. & Edwards, P.J. (eds), Colonization, Succession and Stability. Blackwell Scientific Publications, Oxford.

Brown, V.K., Hendrix, S.D. & Dingle, H. 1987. Plants and insects in early old field succession: comparison of an English site and an American site. Biological Journal of the Linnean Society 31: 59-74.

Bryant, J.P. & Chapin, F.S. III 1986. Browsing-woody plant interactions during boreal forest plant succession. pp. 213-225. In: Van Cleve, K., Chapin, F.S. III, Flanagan, P.W., Viereck, L.A. & Dyrness, C.R. (eds), Forest Ecosystems in the Alaskan Taiga. Springer-Verlag, New York,

Crawley, M.J. 1989. The relative importance of vertebrate and invertebrate herbivores in plant population dynamics. pp 45-71. In: Bernays, E.A. (ed),. Insect-Plant Interactions Vol 1. CRC Press Inc., Florida.

Davidson, D.W. 1993. The effects of herbivory and granivory on terrestrial plant succession. Oikos 68: 23-35.

Edwards-Jones, G. 1988. Insect herbivore load in relation to successional status of the habitat. PhD thesis, University of London.

Edwards-Jones, G. & Brown, V.K. 1993. Successional trends in insect herbivore population densities: a field test of a hypothesis. Oikos 66: 463-471.

Egler, F.E. 1954. Vegetation science concepts. I. Initial floristic composition - a factor in old-field vegetation development. Vegetatio 4: 412-417.

Gange, A.C. 1990. Effects of insect herbivory on herbaceous plants. pp. 49-62. In: Burdon, J.J. & Leather, S.R. (eds), Pests, Pathogens and Plant Communities. Blackwell Scientific Publications, Oxford.

Gange, A.C., Brown, V.K. & Farmer, L.M. 1990. A test of mycorrhizal benefit in an early successional plant community. New Phytologist 115: 85-91.

Gange, A.C., Brown, V.K. & Farmer, L.M. 1992. Effects of pesticides on the germination of weed seeds: implications for manipulative experiments. Journal of Applied Ecology 29: 303-310.

Gibson, C.W. D., Brown, V.K. & Jepsen, M. 1987. Relationships between the effects of insect herbivory and sheep grazing on seasonal changes in an early successional plant community. Oecologia 71: 245-253.

Gibson, C.W.D. & Brown, V.K. 1991. The nature and rate of development of calcareous grassland in Southern Britain. Biological Conservation 58: 297-316.

Hendrix, S.D., Brown, V.K. & Gange, A.C. 1988. Effects of insect herbivory on early plant succession: comparison of an English site and an American site. Biological Journal of the Linnean Society 35: 205-216.

Huntly, N. 1991. Herbivores and the dynamics of communities and ecosystems. Annual Review of Ecology and Systematics 22: 477-503.

Luken, J.O. 1990. Directing Ecological Succession. Chapman & Hall, London.

McBrien, H., Harmsen, R. & Crowder, A. 1983. A case of insect grazing affecting plant succession. Ecology 64: 1035-1039.

McEvoy, P.B., Rudd, N.T., Cox, C.T. & Huso, M. 1993. Disturbance, competition and herbivory effects on Ragwort *Senecio jacobaea* populations. Ecological Monographs 63: 55-75.

Root, R.B. 1996. Herbivore pressure on goldenrods (*Solidago altissima*): its variation and cumulative effects. Ecology 77: 1074-1087.

Southwood, T.R. E., Brown, V.K. & Reader, P.M. 1979. The relationships of plant and insect densities in succession. Biological Journal of the Linnean Society 12: 327-348.

Stinson, C.S.A. 1983. Effects of insect herbivores on early successional habitats. PhD thesis, University of London.

Watt, A.S. 1981. A comparison of grazed and ungrazed grassland A in East Anglian Breckland. Journal of Ecology 69: 499-508.

Wilson, A.D. & Shure, D.J. 1993. Plant competition and nutrient limitation during early succession in the Southern Appalachian mountains. American Midland Naturalist 129: 1-9.

THE IMPORTANCE OF BIOGEOGRAPHY TO BIODIVERSITY OF BIRD COMMUNITIES OF CONIFEROUS FORESTS

HERMANN MATTES

Institute of Landscape Ecology, Robert Koch-Str. 26, D-48149 Münster, Germany

Key words: boreal bird species, saturation of biocoenosis

Abstract

Bird communities of coniferous forests attain their largest species numbers in central Siberia; these numbers are gradually declining towards the periphery of the boreal zone. A secondary maximum of species numbers is found in the Alps. Boreal belts of southern (Mediterranean) mountains are impoverished in numbers of species, although "xenoecious" species inhabit coniferous forests. The uneven geographical dispersion of species is caused by isolation and migration paths during Pleistocene changes of vegetation belts. Considering the subspecies level, Siberian elements reach the Alps and the Rhodope Mountains, Palaearctic elements even extend to the Canaries. Mountain regions along the Mediterranean sea are characterised by many endemic subspecies. Boreal ecosystems include the most specialised species. Similarity of bird fauna assemblages is highest between boreal regions and lowest in Mediterranean areas. Biodiversity diminishes in the boreal forests from central Siberia onwards, in southern extrazonal regions from the Alps to periphery regions. Small isolated populations on southern mountain tops are most endangered.

Introduction

Biodiversity depends on the gene pool of a region, and is specially high if ecological niches are densely packed. Building up a well-balanced community takes time. However, boreal ecosystems were newly established in the post-glacial period. There is only one exception, the "light taiga" in eastern Siberia (Nasarenko 1989). It survived the last glacial period near Lake Baikal. The boreal ecosystems now extend over a large zonal belt, and there are great numbers of extrazonal outposts of boreal communities in southern mountains. Boreal species have been strongly selected by their capability of migrating or dislocating. The numbers of boreal species in glacial refuges should be well differentiated by the distances to nearest refuges.

In the avifauna the importance of biogeography can be clearly demonstrated. Taxonomy, distribution, and ecology of birds are well-known. Despite high mobility

A. Kratochwil (ed.), Biodiversity in ecosystems, 147-156

they show very often a distinct fidelity to one place. Also migrating birds mostly return to the same locality due to their enormous capability of navigation. Birds therefore rarely colonise new areas by chance. Normally there is a continuous expansion of range.

Already Stegmann (1938) had shown a gradually diminishing number of boreal bird species from Siberia to Europe. On the other hand Johansen (1954) postulated a faunal border at the Yenisei valley between west and east Palaearctic realms. Several questions arise as to the biogeography of the boreal ecosystems. Can the boreal zone be looked upon as homogeneous? How strong is the connection between the core areas and their outposts? What influences have surrounding ecosystems? The avifauna might give a special sight of things, but will at least contribute some interesting aspects of recent chorology.

Methods

Bird species are considered as typical boreal if they regularly reproduce in the taiga and/or in the subalpine coniferous forest belts of mountains, respectively. Subspecies of birds were effectively used for comparisons of regions. Nomenclature and subspecific classification follow Bauer & Glutz von Blotzheim (1966 ff.) and Glutz von Blotzheim & Bauer (1971 ff.); for species not yet treated in this handbook Vaurie (1959) is conclusive.

Week clines are not considered to be valuable subspecies. Furthermore, bird species were excluded if their occurrence mainly depended on habitat structures outside the forest. This concerns water fowl (*e.g. Mergus albellus* L., *Bucephala clangula* [L.]), birds on rocks (*Phoenicurus ochruros* [Gm.]), and birds feeding mainly outside coniferous forests (*Columba palumbus* L., *Jynx torquilla* L., *Sturnus vulgaris* L., *Carduelis cannabina* [L.]). Southern taiga regions and montane belts have been omitted because there often is an intensive contact with communities of deciduous forests.

Out of nineteen regions, six represent the boreal zone, thirteen are more or less isolated mountainous areas in the west Palaearctic. The breeding bird populations of coniferous forests of each region were compiled after handbooks, own observations, and special references (see appendix).

Similarity of bird species diversity was calculated by the index of Sörensen ($I = 2\ C/A + B$, C = number of species common to both areas, A and B = total numbers of bird species of each area).

Faunal elements integrate (sub)species originating from the same (hypothetical) refugial areas. As Palaearctic elements subspecies of a huge range from eastern Siberia to western Europe were summarised; Siberian elements originate from northern Asia, European elements from the Balkans, the Alps, or Pyrenees; Mediterranean elements comprise species with a more southern range near the Mediterranean sea and from nearby mountain chains; endemic elements are restricted to isolated populations or small ranges; Turkestanian and Mongolian elements summarise species of Central Asia. For Siberian regions, Siberian elements have been separated into western and eastern species. They probably survived in refuges near the Altai Mountains or near Lake Baikal, respectively, or in Manchuria.

Bird species have been divided into three ecological groups. As "stenoecious" I considered species with a clear preference for coniferous habitats; but they may not necessarily completely avoid other forest types. "Euryoecious" species also live in coniferous habitats as well as in other types of habitats, not showing a clear preference. "Xenoecious" shall be species which throughout most of their ranges prefer other habitats to coniferous forests, but under special conditions they may populate or even prefer coniferous habitats; this occurs mostly in isolated areas.

Results

The central Siberian regions (Yenisei middle taiga, Altai Mountains, Transbaikalia) show the highest numbers of bird species living in coniferous forests (Fig. 1 and 2). From here the species numbers are gradually diminishing to the more eastern (Yakutia) and western regions (Petchora basin, Lapland), but are still higher than in any other region outside the true boreal zone.

A secondary maximum of bird species is found in the Alps and adjacent high mountains. Also, there are many species living in coniferous forests in the Caucasian Mountains. All other regions considered have smaller numbers of boreal bird species, the smallest originating from the archipelago of the Canary Islands which is extremely isolated.

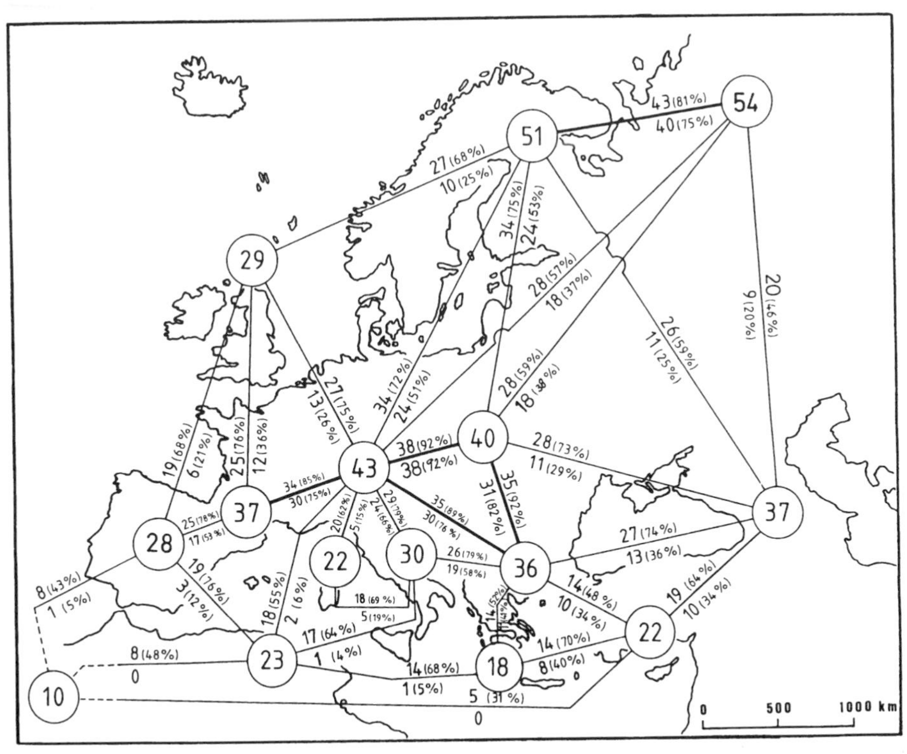

Figure 1 and 2. Numbers of bird species in coniferous forests of European and Siberian regions; numbers of species (upper lines) and numbers of subspecies (lower lines) in common; in brackets similarity index calculated after Sörensen.

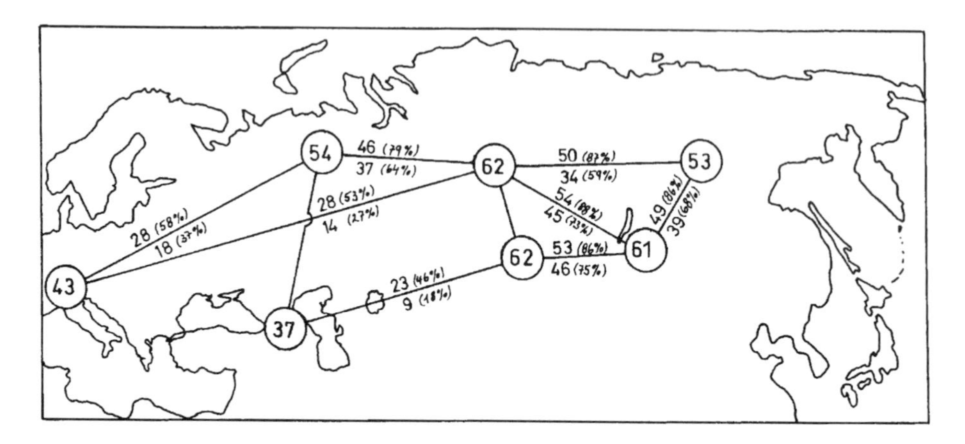

Figure 1 and 2. Numbers of bird species in coniferous forests of European and Siberian regions; numbers of species (upper lines) and numbers of subspecies (lower lines) in common; in brackets similarity index calculated after Sörensen.

The similarities of the coniferous forest faunas are high throughout the boreal zone, especially within central Siberia. Also, there is a high similarity between the faunas of the Alps and of adjacent high mountains. The Caucasian area is nearer to European regions than to the northern ones as to species, but is clearly separated on the subspecies level. Mediterranean regions and Scotland show little similarity to any other region, particularly on the subspecies level.

Considering faunal elements (Fig. 3 and 4) the bird fauna of the coniferous forests mainly consists of Siberian elements in the northern zone, of European elements in Central Europe and in the Balkan peninsula, and of endemic elements in the Mediterranean regions. Mediterranean faunal elements are most abundant in the Iberian peninsula and in Asia Minor.

The ecological preferences of the bird species (Table 1) show clear differences. Corresponding to the total number of bird species, stenoecious species are most numerous in central Siberia. Their numbers are still large in the Alps and adjacent mountains. In isolated regions the share of xenoecious species is relatively high.

Discussion

Without doubt species numbers and ecological adaptations of species to coniferous forests are highest in central Siberia. But this cannot conclusively be explained by recent ecological conditions. The climate is extremely severe with low winter temperatures, and a short breeding season in late spring and early summer. Migration pathways either have to cross the large Central Asian steppes and deserts, or must include long detours via eastern Europe or eastern Asia, respectively. Vegetation is not very luxurious, and rather similar in all boreal ecosystems. Five coniferous tree species form the taiga forest in central Siberia and the subalpine belt in the Alps (Table 1); northern Scandinavian forests only consist of two coniferous tree species. There is no clear relation between species numbers of coniferous trees and species numbers of birds. This also holds true if we consider structural types of coniferous trees instead of species numbers.

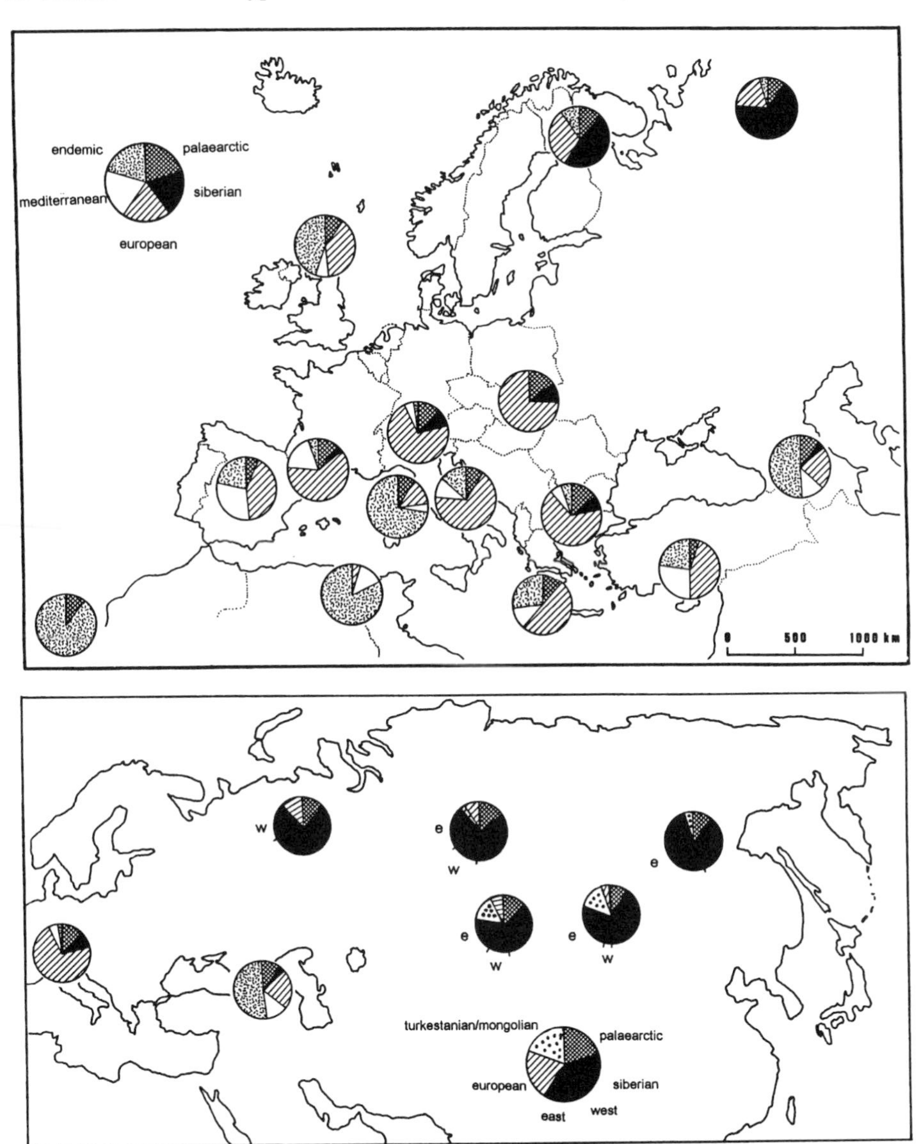

Figure 3 and 4. Faunal elements in European and Siberian regions.

The comparison between the boreal zone and subalpine belts is complicated by different area sizes. Species numbers were taken from areas of about the same size (approx. 100,000 km^2) as far as it was possible. But some of the southern mountain regions are smaller than this. Moreover, northern regions are surrounded by large areas of similar communities. On the other hand mountain areas have much more diverse sites. Finally we register regions of similar size (Alps, Lapland, Altai Mountains) with very different species numbers, and small regions (Rhodope Mountains, Tatra Mountains) with higher species numbers than large areas (Atlas and Taurus Mountains).

It may also be difficult to judge whether a species truly belongs to boreal communities. Data of species and regions are not homogeneous. In some cases it was not clear what importance birch and other deciduous trees have for some bird species to maintain their populations in coniferous forests.

In spite of some uncertain components we can confirm that recent ecological traits do not sufficiently explain numbers of bird species in boreal regions. However, if we consider the geographical position of the refuge areas, bird species numbers differ in accordance with distances between refuges and recent boreal regions. Boreal vegetation was able to survive the last glacial period mainly in the South of the Altai Mountains ("dark taiga" of "Urman" type), near Lake Baikal ("light taiga"), around the Mediterranean Sea, and in the southern Urals (Nasarenko 1989; Frenzel 1968). We can assume that most faunal elements of the boreal ecosystems were associated with boreal vegetation.

Table 1. Ecological groups of bird species, coniferous tree species, and numbers of structural types of coniferous trees (types: *Picea/Abies*; *Larix*; *Pinus* (haploxylon); *Pinus* (diploxylon); *Cedrus*)

area	bird species				species of coniferous trees	types of tree structures
	total (n)	% euryoe-cious	% stenoe-cious	% xenoe-cious		
Central Siberia (Yenisei)	62	60	40	0	*Picea obovata, Abies sibirica, Larix sibirica, Pinus sibirica, P. sylvestris*	4
Yakutia	53	53	47	0	*Larix daurica, Pinus pumila*	2
Lapland	51	61	39	0	*Picea abies, Pinus sylvestris*	2
Alps	43	65	33	2	*Picea abies, Larix decidua, Pinus cembra, P. sylvestris, P. mugo*	4
Rhodope Mts.	36	64	36	0	*Picea alba, Abies alba, Pinus sylvestris*	2
Pyrenees	37	65	27	8	*Abies alba, Pinus sylvestris, Pinus uncinata*	2
Caucasus	37	76	19	5	*Picea orientalis, Abies nordmanniana, Pinus sylvestris*	2
Taurus	22	64	18	18	*Abies cilicica, Pinus nigra, Cedrus libani*	3
Central Spain	28	68	21	11	*Pinus sylvestris, Pinus nigra*	1
Atlas Mts.	23	70	17	13	*Abies numidica, Cedrus atlantica*	2
Tenerife	10	70	20	10	*Pinus canariensis*	1

From west and east Siberian refuge areas, the distances to central Siberia were very small. In the Yenisei valley the different faunas from the Altai Mountains and the Baikal region met, and therefore it was called a "faunal border" already by members of early expeditions (Pallas in Ilicev & Flint 1985). Later on it was documented in detail by Johansen (1954) and others. The faunal border presents itself as a relatively sharp separation of west and east Palaearctic faunas because it is also an ecological border line. In the West we find the dark west Siberian taiga and waste moorland. In the East the central Siberian mountains tower; they are predominantly covered by the "light taiga". The Yenisei valley itself is a splendid line of expanding ranges. In consequence the middle taiga region at the Yenisei shows an unusually high biodiversity of bird species (and of many other animal groups, too).

From central Siberia to the periphery of the boreal zone, bird species numbers are gradually decreasing. The bird fauna of the boreal zone shows a greater similarity between any region within the zone than with the southern mountain region. Lapland is subject to the highest influence of European elements, and shows the greatest similarity to southern regions in bird species. Some species, such as *Phylloscopus borealis* (Blasius), *Ph. trochiloides* (Sundevall), *Turdus pilaris* (L.), and others have expanded their ranges only recently. The process of expanding from refuge areas seems not yet completed.

The bird faunas of the Pyrenees, the Alps, the Carpathians, and the Rhodope Mountains are very similar. Many identical subspecies indicate an extensive exchange between these mountain areas. There is a striking concordance of bird fauna with vegetation and type of altitudinal belts. Ozenda (1985) postulated an alpine type of altitudinal belts which comprises apart from the Alps the Pyrenees, the Sudeten and the Carpathian Mountains, the Balkan Mountains (including Rhodope), the Apennines, and Corsica. The Apennines and Corsica have impoverished bird faunas; the species left are nearly identical with those of the Alps. In contrast, the Mediterranean regions differ considerably from each other.

The Caucasus is faunistically isolated. The number of species is high, but there exist many endemic subspecies. Most accordance we find with the Rhodope Mountains, less with northern and eastern regions. In the Mediterranean regions endemic forms also developed, however the bird faunas are impoverished. Few species which are common to most regions derive from a stock of Palaearctic species, consisting *e.g.* of *Scolopax rusticola* L., *Cuculus canorus* L., *Parus major* L., *P. ater* L., and *Loxia curvirostra* L. which belong to the same subspecies over large areas.

Cluster analysis (dendrogram in Fig. 5) supports the clear separation of northern boreal bird faunas from the extrazonal southern faunas as well as the expressed similarity of the faunas of the Alps and neighbouring areas. A discriminate analysis on the level of seven groups (arrow in Fig. 5) verifies the clusters, except the Apennines (should be single) and the Transbaikal region (better combined with Yakutia).

It is interesting that regions with reduced boreal bird faunas typically have "xenoecious" species in coniferous forests. As "xenoecious" I would like to define species which over a wide part of their range inhabit deciduous forests only, but in isolated areas penetrate into "foreign" coniferous forests. It is assumed that they successfully compete in coniferous habitats because truly adapted species are nearly completely lacking. Examples of "xenoecious" species in southern coniferous forests are *Certhia brachydactyla* Brehm, *Parus caeruleus* L., *P. lugubris* Temminck, and *Picus viridis* L..

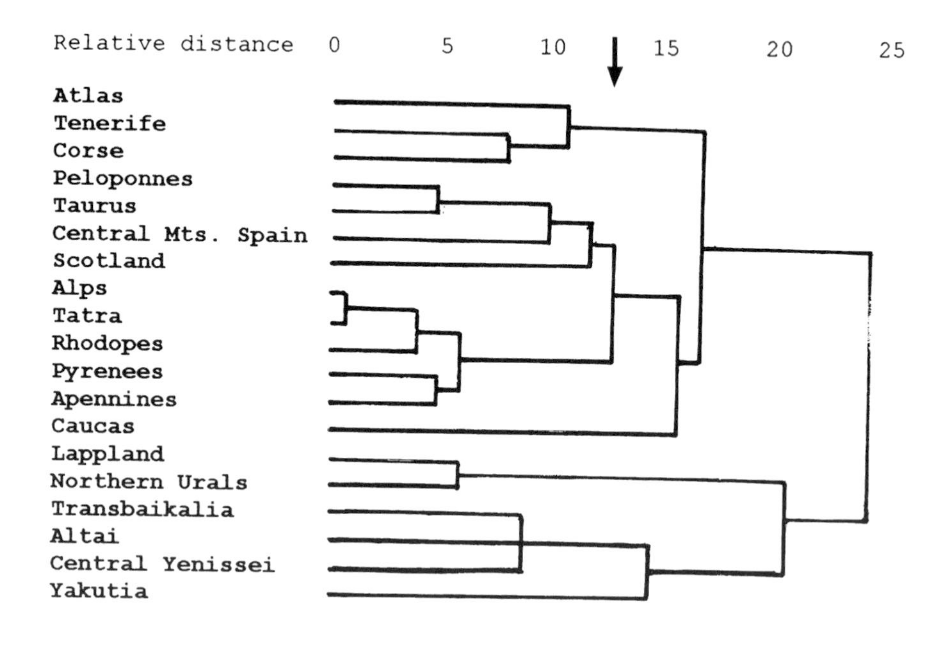

Figure 5. Dendrogram from a cluster analysis using complete linkage for agglomeration of subspecies as variables. The arrow indicates the level of 7 clusters.

Typical boreal species specialised on coniferous forests, *e.g. Glaucidium passerinum* (L.), *Picoides tridactylus* (L.), *Regulus regulus* (L.), *Parus ater, Nucifraga caryocatactes* (L.), and others mostly reach the border of their range in the Alps and Rhodope Mountains. It corresponds to the southern border of *Picea abies*. Few Palaearctic faunal elements expanded further south, *Scolopax rusticola* reaching even the Canary Islands. It is the question whether the ecosystems in the Mediterranean mountains can really be called "boreal". Among the Mediterranean bird fauna there are only very few species which are specially adapted to high-altitude coniferous forests. Examples are *Serinus citrinella* (Pallas) and, with certain reservations, *Serinus pusillus.* Typically Mediterranean mountain bird species prefer very open, stony habitats with scattered bushes or trees. In consequence more northern euryoecious species inhabit the mountain forests. They form an impoverished fauna which still can be called "boreal".

Bird faunas of extrazonal coniferous forests show reduced biodiversity, caused by isolated areas and/or large distances to genetic resources during Pleistocene history. It is supposed that the biocoenoses are not saturated, probably they could involve some more species apart from the "xenoecious" ones. Yet the impoverished and isolated assemblages of birds might be stable because of a high share of euryoecious species.

Some of the endemic species are seriously endangered because of loss of habitats. This is specially the case for *Sitta ledanti* (Vieillard) (Gatter & Mattes 1979) and

Fringilla teydea (Webb, B. & M.) which live in very restricted habitats. Boreal ecosystems are protected by their large ranges but are susceptible to many human or natural influences. Therefore it is very important that Russia has established various large biosphere reservations in several taiga regions.

References

Bauer, K.M. & Glutz von Blotzheim, U.N. 1966 ff. Handbuch der Vögel Mitteleuropas. Wiesbaden.
Brichetti, P. 1978. Guida della uccelli nidificanti in Italia. Brescia.
Cramp, S. 1977 ff. Handbook of the birds of Europe, the Middle East and North Africa. Oxford.
Feriancova-Masarova, Z. & Dontschew, S. 1969/70. Gegenüberstellung der Ornithofauna der Westkarpaten (Slowakei) und der westlichen Stara Planina (Bulgarien). Biologica 24 und 25: 356-373 und 85-101.
Flint, V.E., Boehme, R.L., Kostin, Y.V. & Kuznetsov, A.A. 1984. A field guide to the birds of the USSR. Princeton.
Frenzel, B. 1968. Grundzüge der pleistozänen Vegetationsgeschichte Nord-Eurasiens. Wiesbaden.
Gatter, W. & Mattes, H. 1979. Zur Populationsgröße und Ökologie des neuentdeckten Kabylenkleibers *Sitta ledanti* Vieillard 1976. Journal für Ornithologie 120: 390-405.
Glutz von Blotzheim, U.N. 1964. Die Brutvögel der Schweiz. Aarau.
Glutz von Blotzheim, U.N. & Bauer, K.M. 1971 ff. Handbuch der Vögel Mitteleuropas. Wiesbaden.
Ilicev, V.D. & Flint, V.E. 1985 ff. Handbuch der Vögel der Sowjetunion. Wiesbaden.
Järvinen, O. & Väisänen, R. 1973. Species diversity of Finnish birds, I: Zoogeographical zonation based on land birds. Ornis Fennica 50: 93-125.
Johansen, H. 1954. Die Jenissej-Faunenscheide in Sibirien und ihr Durchbruch. Acta Congr. Internat. Ornith. 11: 383-386.
Kasparek, M. 1992. Die Vögel der Türkei - eine Übersicht. München.
Mattes, H. 1988. Untersuchungen zur Ökologie und Biogeographie der Vogelgemeinschaften des Lärchen-Arvenwaldes im Engadin. Münstersche Geographische Arbeiten 30: 1-138.
Matvejev, S.D. 1976. Survey of the Balkan peninsula bird fauna. Serb. Acad. Sci. Arts Monography No. 46 (Belgrad).
Nazarenko, A. 1989. Transzonal interchange of faunal elements between south and north Asia as the typical event of the newest, late Pleistocene-Holocene, history of Palaearctic dendrophilous avifauna. Verhandlungen der Deutschen Ornithologischen Gesellschaft Sonderheft.
Ozenda, P. 1985. La végétation de la caine alpine dans l'espace montagnard européen. Masson, Paris.
Purroy, F.J. 1972. Communidades de aves nidificantes en el bosque pirenaico de abeto blanco (*Abies alba* L.). Boletin de Estacion central de Ecologia 1: 41-44.
Purroy, F.J. 1974. Contribucion al conocimiento ornithologico de los pinares pirenaicos. Ardeola 20:245-261.
Rogacheva, H. 1992. The birds of central Siberia. Husum.
Sharrock, J.T.R. 1976. The atlas of breeding birds in Britain and Ireland. Brit. Trust Ornith. and Irish Wildbird Conserv., Tring.
Stegmann, B. 1932. Die Herkunft der paläarktischen Taiga-Vögel. Archiv für Naturgeschichte, Neue Folge 1: 355-398.
Stein, J. 1976. Die Schutzwürdigkeit des Urwaldgebietes Kentriki Rodopi (Griechenland). IUCN, London.
Telleria, J.L. 1980. The bird communities of Scots pine (*P. sylvestris* L.) forests of the sistema central mountains (Spain) censused by means of the frequential sampling method. pp. 221-225. In: Oelke, H. (ed), Bird census work and nature conservation. Göttingen.
Thibault, J.-C. 1983. Les Oiseaux de la Corse. Ajaccio.
Vaurie, C. 1959. The birds of the Palaearctic fauna. Passeriformes. London.
Voous, K.H. 1962. Die Vogelwelt Europas. Hamburg/Berlin.

Appendix

Bird faunas of 19 regions were compiled from Bauer & Glutz von Blotzheim (1966 ff.), Glutz von Blotzheim & Bauer (1971 ff.), Cramp (1977 ff.), Ilicev & Flint (1985 ff.), Voous (1962), and own observations (Mattes 1988).
Special references are mentioned for each region. The regions considered are *in the boreal zone:*

Lapland (Järvinen & Väisanen 1973)
Upper Petchora basin (Flint *et al.* 1984)
Yakutia (Flint *et al.* 1984)
Transbaikal region (Flint *et al.* 1984)
Mid taiga of central Siberia (Rogacheva 1992)
Altai Mountains (Rogacheva 1992; Flint *et al.* 1984);

in southern subalpine belts:

Central Alps (Glutz 1964; Mattes 1988)
Rhodope (Stein 1976; Matvejev 1976)
Etruscian and central Apennines (Brichetti 1976)
Aures and Atlas Mts.
Tenerife
Corsica (Thibault 1983)
Tatra (Feriancova-Masarova & Dontschew 1969)
Caucasia
Central Taurus (Kasparek 1992)
Peloponnes
Pyrenees (Purroy 1972 & 1974)
Sierras de Guadarrama y de Gredos (central Spain) (Telleria 1980)
Scottish Highlands (Sharrock 1976)

LIMITS AND CONSERVATION OF THE SPECIES DIVERSITY OF SMALL MAMMAL COMMUNITIES

RÜDIGER SCHRÖPFER

Fachgebiet Ethologie, FB Biologie/Chemie, Universität Osnabrück, Barbarastr. 11, 49069 Osnabrück, Germany

Key words: small mammal communities, species diversity

Introduction

It has become necessary and usual to evaluate communities and to illustrate these values by different indices. For that, among other things, the Shannon-index and the evenness E are more and more used.

Although the common textbooks for example specially point out that the index, concerning the number of species, the number of individuals and especially their distribution among the species, has to be examined carefully, it is applied rather casually (Begon *et al.* 1986).

The problems become clear when dealing with communities for which only a few dominant species are characteristic. The H'-index is low due to the varying frequencies of individuals per species, since high H'- or E-values indicate a low dominance in the community. So, in an evaluation according to H' and E, communities in which the individuals are distributed proportionally rate higher than those in which the distribution is unproportional. That means that only communities which are either dominant or show a proportional distribution in their structure can be compared with each other in this way. Such an approach shall be applied to small mammal communities with shrews and rodent species.

The qualitative structure

According to Schröpfer (1990) epigeal communities of small mammals are composed of guilds. In Europe, the guilds are structured corresponding to the families, *e.g.* Soricidae, Muridae, Arvicolidae. This does not only seem to agree with the different trophic preferences but also with the different types of locomotion (Schröpfer & Hake 1990). So far, up to 3 species have been found in each guild; they differ in size (*e.g.* head-body length) at least by the factors 1.3 to 1.5 (Fig. 1). Vicariances are the basis of this guild structure: mice (*e.g. Apodemus sylvaticus*/*Apodemus flavicollis*) and voles (*e.g. Microtus agrestis*/*Microtus oeconomus*) of the same size exclude one another. Trophic demands and foraging strategies seem to be very similar in similar species; that has to be studied in detail (Schröpfer in prep.).

A. Kratochwil (ed.), Biodiversity in ecosystems, 157-162

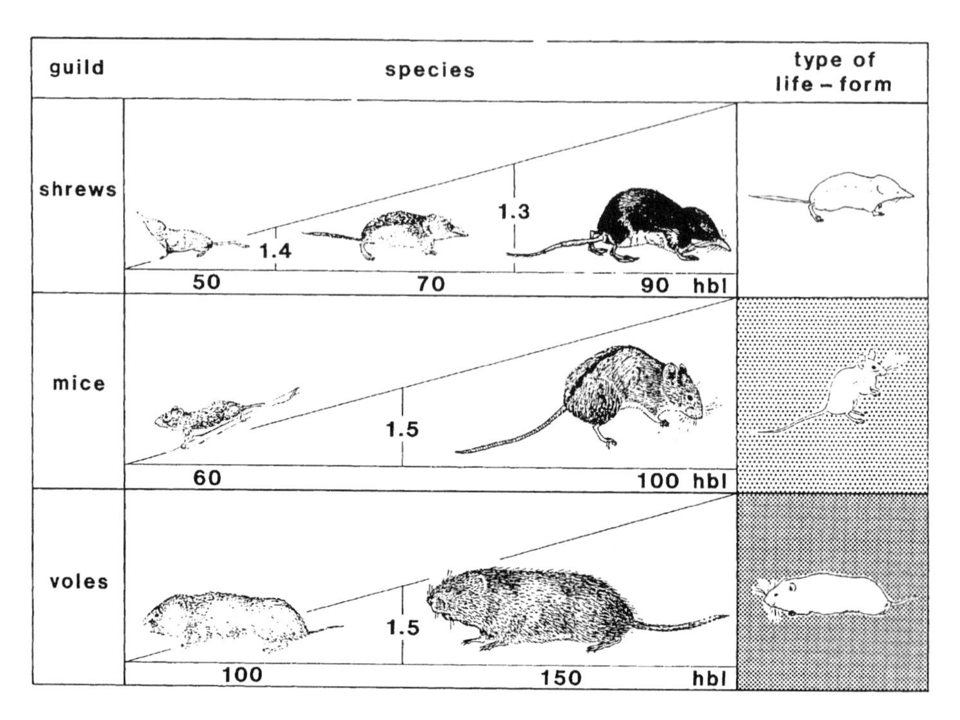

Figure 1. Small mammal community in a reed (*Phragmitetum*) of Central Europe. Guild of shrews: *Sorex minutus, Sorex araneus, Neomys fodiens*; guild of mice: *Micromys minutus, Apodemus agrarius*; guild of voles: *Microtus agrestis, Arvicola terrestris.* Type of life-form: shrew-type, mouse-type, vole-type; hbl: head-body length in mm; 1: the hbl-ratio between the neighbouring species.

According to this model, three guilds at the most, each with three species, can be expected in epigeal small mammal communities. That means that maximally nine species can form such a community (shrews, rodent species).

The quantitative structure

Looking at the existing results, the individuals in a community are not distributed proportionally among the species (Schröpfer 1989). Usually over 75% of the individuals belong to three species; these species are called principal species (Schröpfer 1990; Fig. 2). The accompanying species, having a strong habitat preference, but occurring only in small numbers, can be found in the remaining part; as well as the external species comprising those individuals which occur in the habitat for example during a dismigration, but have their demotopes in other habitat types on account of a different habitat preference (Fig. 3).

Usually, each principal species belongs to another guild. As the principal species are named after the method of inverse rank, it can happen that two of the three species come from the same guild, for example rodent species in gradation years. Then these two species differ in body size (see above chapter).

These results show that the proportion of the individuals per species in a small mammal community is strongly asymmetric

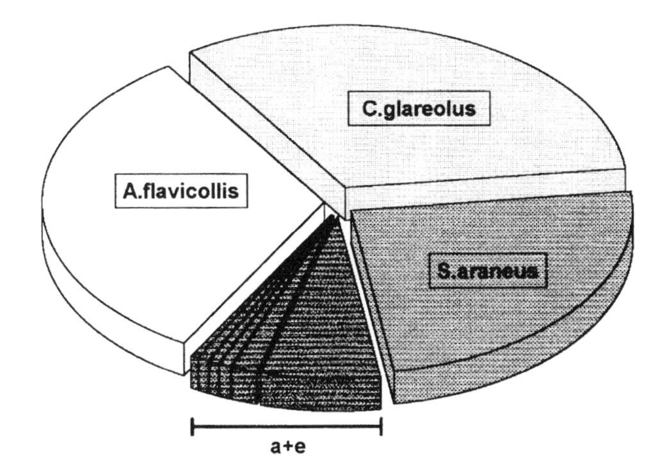

Figure 2. Proportion of the individuals of the three principal species, the yellow-necked mouse (*A. flavicollis*), the bank vole (*C. glareolus*) and the common shrew (*S. araneus*) in the small mammal community of the woodland habitats in the International Park "Unteres Odertal"/East Germany. a, e: see Figure 3.

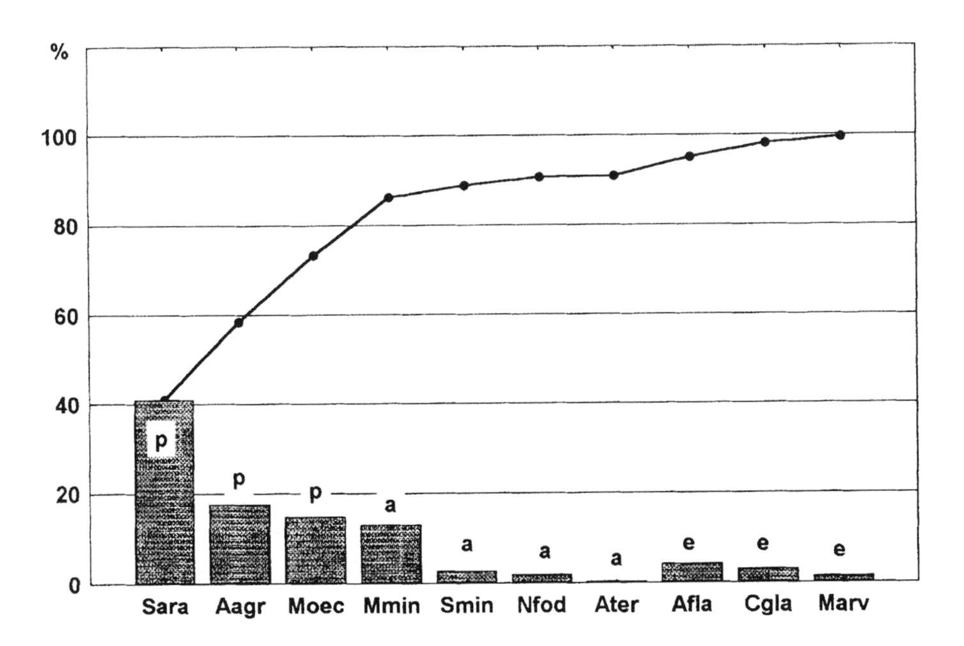

Figure 3. The species proportion and the cumulative addition of the individuals of ten small mammal species in the reed habitats of the International Park "Unteres Odertal"/East Germany. p: principal species, a: accompanying species, e: external species. Sara: Sorex araneus, Aagr: Apodemus agrarius, Moec: Microtus oeconomus, Mmin: *Micromys minutus*, Smin: *Sorex minutus*, Nfod: *Neomys fodiens*, Ater: *Arvicola terrestris*, Afla: *Apodemus flavicollis*, Cgl: *Clethrionomys glareolus*, Marv: *Microtus arvalis*.

The limits of species diversity

Such a structure of a small mammal community influences the size of the diversity indices. This is always the case if the number of species and the number of individuals per species are included in the calculation. An increase in the number of species enlarges the index, whereas a strong asymmetry in the distribution lowers it. When a calculation is done for all species given in Fig. 3, that is for the principal species (p), the accompanying species (a) and the external species (e), the result is H' 1,7 (species n=10; individuals n=1.175). If the external species are not taken into account, since they are not in their own demotope, as specified above, the value of H' is 1,4 (species n=7; individuals n=1.074). That means that only the syntopic species should be included in the calculation of species diversity. According to the model, only the principal and the accompanying species are syntopic species and can amount to the maximal number of species. For that reason the Shannon-indices of small mammal communities of European habitats are relatively low (Fig. 4). The highest can be found in *Alnetum* habitats and in reed (*Phragmitetum*). The high value of the ecotones is caused by the edge effect. For most small mammals ecotones represent a fighting zone with high interference.

Fig. 5 shows clearly how similar the distribution of the individuals in the communities is in each habitat. The values of evenness gather around 0,7. This shows that a very similar distribution of individuals was found in all habitats in which the principal species form the greatest part.

The species diversity in epigeal small mammal communities is limited because of the qualitative and quantitative structure, it cannot be expected in any size.

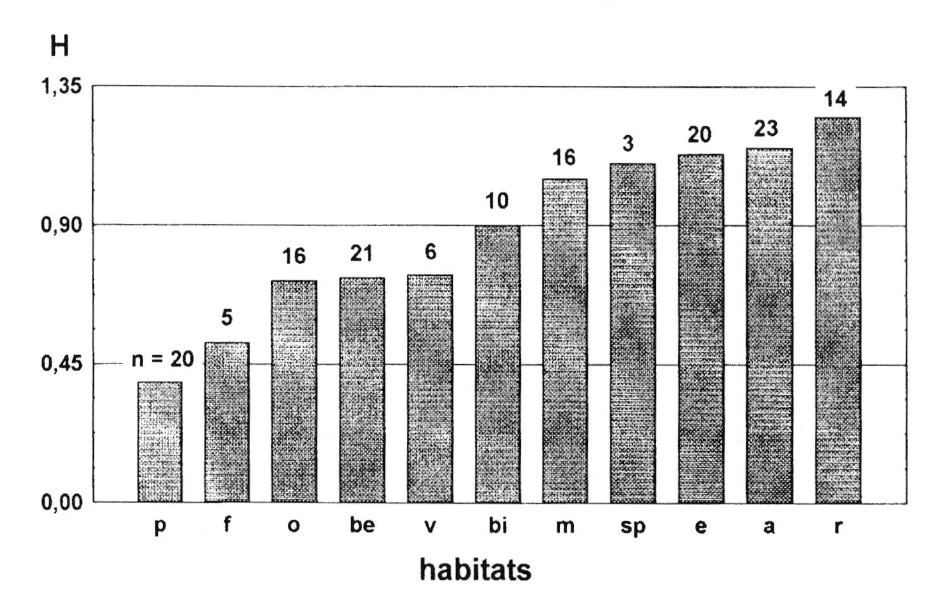

Figure 4. Shannon-index (mean values) of small mammal communities in different habitats (shrews and rodent species). p: pasture, f: field, o: oak-hornbeam forest, be: beech forest, v: village, bi: birch forest, m: meadow, sp: spruce forest, e: ecotone, a: alder forest, r: reed. H': Shannon-index, n: number of each habitat type.

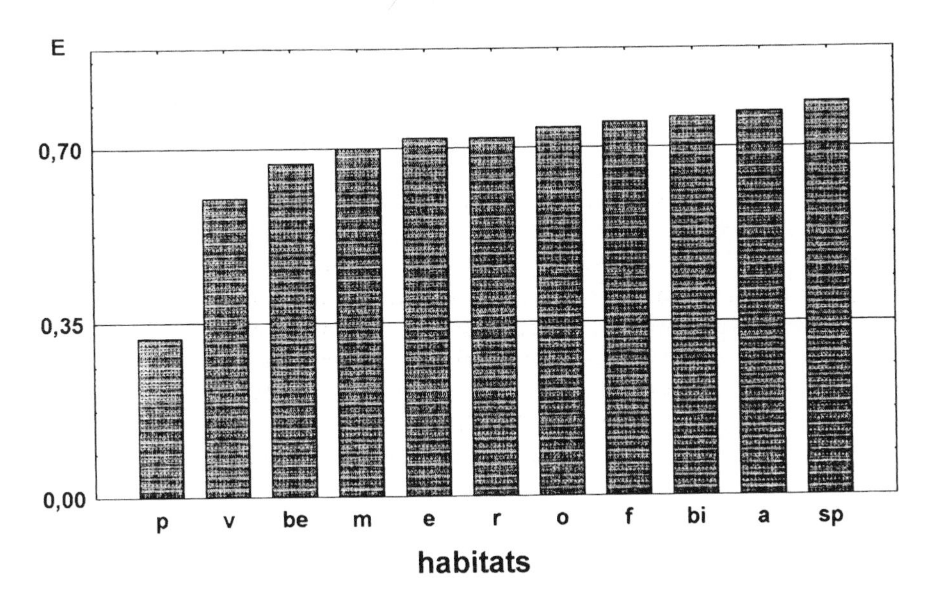

Figure 5. Evenness (mean values) of small mammal communities in different habitats (shrews and rodent species); legend see Figure 4.

The conservation

The above comments show that the size of the diversity index can only partly be used to justify that a small mammal community is worth being conserved (Gaisler & Nesvadbova 1994). Popular effects (like gradations), the habitat preference (like the collection of requisites), the interspecific competition (like vicariances) and the edge effect (like species interferences) influence the number of resident species and the distribution of individuals very strongly. Only those species for which the habitat is their demotope should be considered for the calculation of the habitat quality and for the determination of the status of the small mammal community. Only the syntopic living resident species form the community (Schröpfer 1992).

With regard to the conservation of a community from this point of view, firstly, the density of the occupation of niches is considered, that means that the capacity of the habitat is calculated; secondly, the species combination typical of the area is seen, that means that the habitat preference of each species is taken into account; and thirdly, the temporality of the popular action is included, that means that the coexistence and interactions of the generation are planned.

That is why syntopic populations which form the community have to be conserved.

References

Begon, M., Harper, J.L. & Townsend, C.R. 1986. Ecology - Individuals, Populations and Communities. Blackwell Sc. Publ., Oxford.

Gaisler, J. & Nesvadbova, J. 1994. Diversität von Kleinsäugern (Insectivora, Chiroptera, Rodentia) in einem UNESCO-Bioreservat. Z. Säugetierkunde, Suppl.: 13-14.

Schröpfer, R. 1989. Habitatpräferenz und Struktur silvicoler Theriozönosen. Verh. Ges. Ökologie 17: 437-443.

Schröpfer, R. 1990. The structure of European small mammal communities. Zool. Jb. Syst. 117: 335-367.

Schröpfer, R. 1992. Biotopschutzmaßnahmen für Säugetiere im Dümmer-Gebiet. Norddeutsche Naturschutzakademie, NNA-Berichte 5/2: 44-48.

Schröpfer, R. & Hake, C. 1990. Ein funktionsmorphologischer Erklärungsversuch für die Nischenbesetzung in Kleinsäugetiergemeinschaften. Z. Säugetierkunde, Suppl.: 43-44.

LEMURS AS INDICATORS FOR ASSESSING BIODIVERSITY IN FOREST ECOSYSTEMS OF MADAGASCAR: WHY IT DOES NOT WORK

JÖRG U. GANZHORN

Zoologisches Institut und Zoologisches Museum, Universität Hamburg, Martin-Luther-King-Platz 3, 20146 Hamburg, Germany

Key words: assembly rule, disturbance, historical biogeography, indicator species, lemurs, Madagascar, nested subsets, primates, species diversity

Abstract

This paper summarizes aspects of the evolution of the present communities of lemurs in Madagascar in view of the usefulness of lemurs as indicators for habitat quality. It is shown here that the number of lemur species present in any given area and the taxonomic and functional composition of their communities follow deterministic patterns. Floristically diverse habitats contain more lemur species than sites with fewer tree species. Thus, the presence of lemur species gives some information about habitat characteristics. This fulfils one of the basic requirements for indicator species, namely that their presence reflects biodiversity in general in a deterministic way. On the other hand, deterministic historical processes eliminated certain lemur species with higher probabilities than others. These past events superimpose the present ecological relationships. They make it impossible to derive unambiguous information on present habitat quality based on the absence of lemur species. Thus, the presence of lemur species gives some information about habitat quality, however, the absence of the same species may not be related to present forest characteristics but be based on habitat characteristics in the past. Therefore, while all processes outlined here are highly deterministic, their combined effects are not easy to predict because of differences in their temporal scales.

Introduction

To unambiguously characterize ecosystems and to define the degree of degradation or recovery from disturbances is one of the prime tasks in community ecology with far-reaching consequences for conservation biology. Especially in humid tropical ecosystems, complexity and species diversity are too high to allow quick characterization and thorough monitoring as required for conservation action. Indicator species may be a suitable substitute for such long-term efforts. Certainly, long-term intensive studies are preferable,

A. Kratochwil (ed.), Biodiversity in ecosystems, 163-174

but often impossible due to constraints of time, money or the need for immediate political decisions. In this context, indicators are defined as organisms "whose characteristics (*e.g.* presence or absence, population density, dispersion, reproductive success) are used as an index of attributes too difficult, inconvenient, or expensive to measure for other species or environmental conditions of interest" (Landres *et al.* 1988).

Due to its large number of endemic plants and animals and the increasing human pressure on its natural ecosystems, Madagascar is one of the countries with the highest conservation priority in the world (Mittermeier *et al.* 1987). This situation requires immediate action, first, to characterize the geographical distribution of whatever is left of Madagascar's different ecosystems (Nelson & Horning 1993a, b; Du Puy & Moat 1996; Smith 1997), second, to define conservation priorities (Conservation International *et al.* unpubl. report; Du Puy & Moat 1996) and third, to allow long-term monitoring of impacts such as human disturbance, natural catastrophes or conservation activities (Kremen *et al.* 1994).

Lemurs might be useful as indicators because they occur in all natural terrestrial habitats of Madagascar, they are ecologically diverse, some of the basic features of their biology are known, they can reach high species diversity (compared to other mammalian taxa), they are easy to identify and a few days' time of survey is sufficient to find all but the most secrete species even in pristine evergreen rain forests (Sterling & Rakotoarison 1998). All of this led to the conclusion that "lemurs are an ideal indicator group for biodiversity planning" (Smith 1997). But, whereas lemurs might be conceptually useful as umbrella species for reserve design in biodiversity *planning*, it has been concluded previously that they are unsuitable for biodiversity *assessment* and *monitoring* (Ganzhorn 1994). This seeming discrepancy between ideas and the distinction between the usefulness of lemurs for planning purposes *versus* their utility for biodiversity assessment calls for further explanation. It will be shown here that and how the assembly and present composition of lemur communities is and has been subject to deterministic processes. But since these deterministic processes operate on different temporal scales, their superposition results in unpredictable scenarios (Remmert 1985). This makes it impossible to derive valid conclusions about habitat biodiversity in general from presence/absence data only.

Taxonomic patterns in lemur species distribution

Biogeographical similarities

The three major vegetation types of Madagascar contain distinct assemblages of lemur species. When looking at similarities of species composition between locations, the sites of the southern spiny forest and the sites of the western dry deciduous forest form a distinct cluster and all the rain forest sites of the east are lumped (Fig. 1). These clusters are based on the exclusive distribution of some species, such as **Lemur catta** in the south, **Indri indri**, **Varecia variegata** and **Eulemur rubriventer** in the east, and small nocturnal genera, like **Cheirogaleus** and **Microcebus** which have distinct sister species in the west and in the east (Martin 1972, 1995; Tattersall 1982; Pollock 1986; Richard & Dewar 1991).

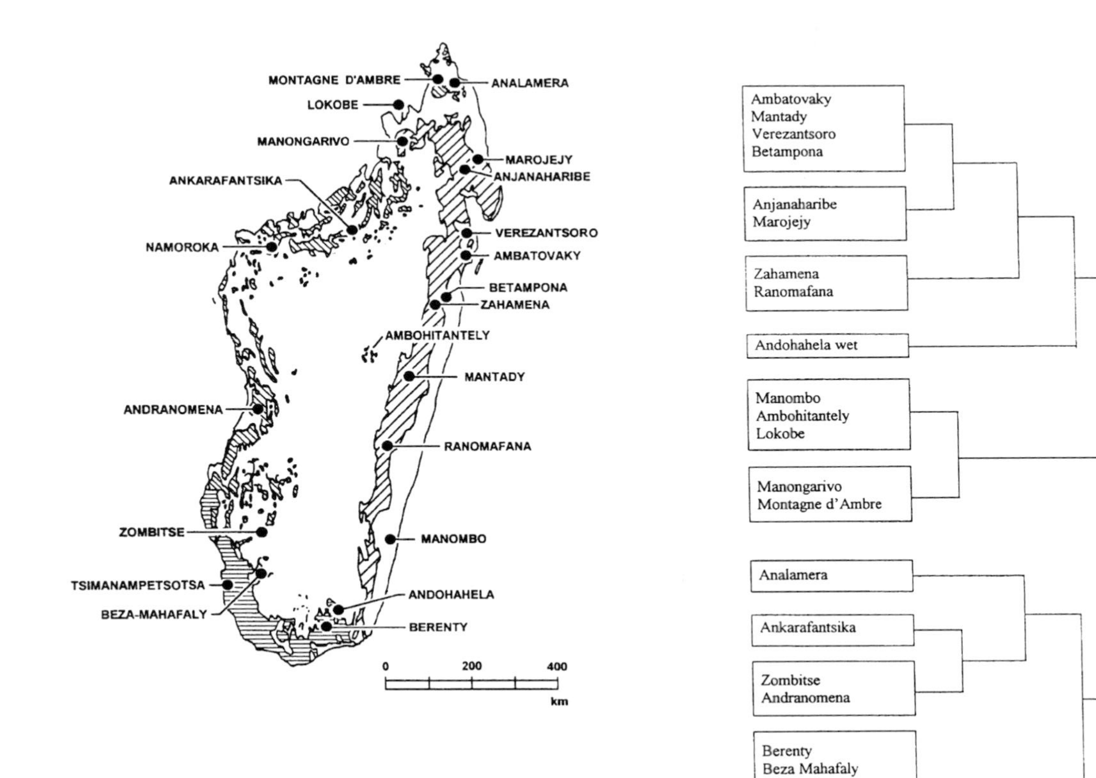

Figure 1. Map of Madagascar showing the distribution of the three major vegetation types (eastern rain forest, western dry deciduous forest, as well as southern dry deciduous and spiny forest), and the sites used in the analysis for similarities in lemur species composition (map of vegetation types and lemur species composition from Nicoll & Langrand 1989, supplemented by Ganzhorn 1994, 1997, 1998).

Nested subsets of species composition

Within the eastern rain forest, the assemblages of true rain forest lemur species found at various sites are composed of almost perfectly nested subsets of species (Table 1; the result of an analysis of the western and southern sites is somewhat different and is interpreted in another publication [Ganzhorn 1998]). In principle, nested patterns of species composition can be based on two different species characteristics: different colonization potential and different vulnerability to extinction, either after habitat modification or due to different requirements on the habitat size to maintain viable populations (Patterson & Atmar 1986). In the first scenario, species originate from a common species pool. From there, they start colonizing the available habitat. Good colonizers will get further than poor colonizers. The result in space and time will be that species assemblages surveyed at different sites will be composed of nested subsets of species, and the species similarities between sites will strongly depend on the distances between sites (Fig. 2).

Table 1. Distribution of rain forest lemur species; from Ganzhorn (1998), extended by new observations by Rakotoarison *et al.* (1996). The distribution differs from a distribution as assembled by the RANDOM1 model of Patterson & Atmar (1986) with $p < 0.001$. *Mr = Microcebus rufus, Ef = Eulemur fulvus, Cm = Cheirogaleus major, Lm = Lepilemur mustelinus, Dm = Daubentonia madagascariensis, Hg = Hapalemur griseus, Al = Avahi laniger, Pd = Propithecus diadema, Ii = Indri indri, Vv = Varecia variegata, Er = Eulemur rubriventer, Hs = Hapalemur simus, Ha = Hapalemur aureus, At = Allocebus trichotis*

	Mr	*Ef*	*Cm*	*Lm*	*Dm*	*Hg*	*Al*	*Pd*	*Ii*	*Vv*	*Er*	*Hs*	*Ha*	*At*	# Species
Zahamena	1	1	1	1	1	1	1	1	1	1	1	1		1	13
Ranomafana	1	1	1	1	1	1	1	1		1	1	1	1		12
Ambatovaky	1	1	1	1	1	1	1	1	1	1	1				11
Mantady	1	1	1	1	1	1	1	1	1	1	1				11
Anjanaharibe	1	1	1	1	1	1	1	1	1		1			1	11
Verezantsoro	1	1	1	1	1	1	1	1	1	1				1	11
Betampona	1	1	1	1	1	1	1	1	1	1					10
Marojejy	1	1	1	1	1	1	1	1			1				9
Andohahela	1	1	1	1	1	1	1	1							8
Manongarivo	1	1	1	1	1	1									6
M.d'Ambre	1	1	1	1	1										5
Manombo	1	1	1			1									4
Ambohitantely	1	1					1								3
Lokobe	1			1											2

Alternatively, all the species in the original species pool occurred over the whole range of the sites in question. But due to local modifications of habitat quality (climatic change, fragmentation, anthropogenic effects), the species most vulnerable to habitat change became extinct first and the others followed one after the other in a defined sequence. The result is also a strongly nested pattern in local species composition, but in contrast to the scenario above, the similarities in species assemblages are uncorrelated with the distances between sites.

For eastern rain forest sites, the similarities between species assemblages of different sites are not related to the geographical distance between them (Spearman rank correlation: r_s = -0.22, p = 0.07; Fig. 3). These two results (highly nested pattern and no distance effect) suggest that the eastern rain forest of Madagascar originally was home to a rather uniform pool of species. Then, changes in habitat quality in the past caused selective but very deterministic extinction of certain lemur species. First, all the larger species became extinct (Richard & Dewar 1991). But obviously local extinction was not only a question of the size of a species. The non-random pattern of species composition in the different extant communities also indicates different extinction probabilities for the different living species. These extinction probabilities seem to be unrelated to body mass. It must be one of the prime tasks of the present studies of lemur biology to identify which characteristics make some of the extant lemur species more vulnerable to extinction than others.

After the postulated Pleistocene subdivision of the originally contiguous belt of the eastern rain forest, the environmental situation changed again and gave rise to the seemingly uniform evergreen rain forest found today along Madagascar's eastern escarpment (for reviews of vegetation changes see MacPhee *et al.* 1985; Burney 1987, 1993). Either natural barriers such as rivers or the lack of time might then have prevented a restitution of the original uniform distribution of all species. This resulted in the present pattern of almost perfectly nested subsets of lemur communities seen today in the rain forests of Madagascar's east coast.

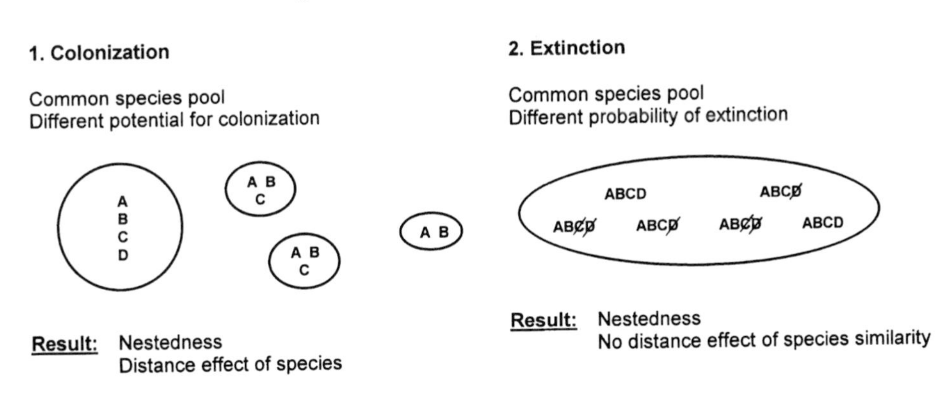

Figure 2. Possible ways to nested patterns in species assemblages: 1. Colonization, 2. Extinction.

Functional patterns in lemur communities

Correlation between numbers of lemur and tree species

A large number of studies revealed rather tight correlations between structural vegetation characteristics, floristic diversity and the number of animal species in various vertebrate taxa (review *e.g.* Putman 1994). For lemur communities the number of lemur species to be found in any given habitat is correlated with the number of tree species present in that area (Fig. 4). Structural diversity seems to be subordinate to floristic diversity on a regional as well as on a national level (Ganzhorn 1994; Ganzhorn *et al.* 1997). The interpretation for this correlation is that, in the phenologically rather seasonal environment of Madagascar's dry and wet forests (Meyers & Wright 1993; Sorg & Rohner 1996), lemurs require a certain number of different tree species to guarantee an all-year supply of fruit and leaves (Ganzhorn *et al.* 1997).

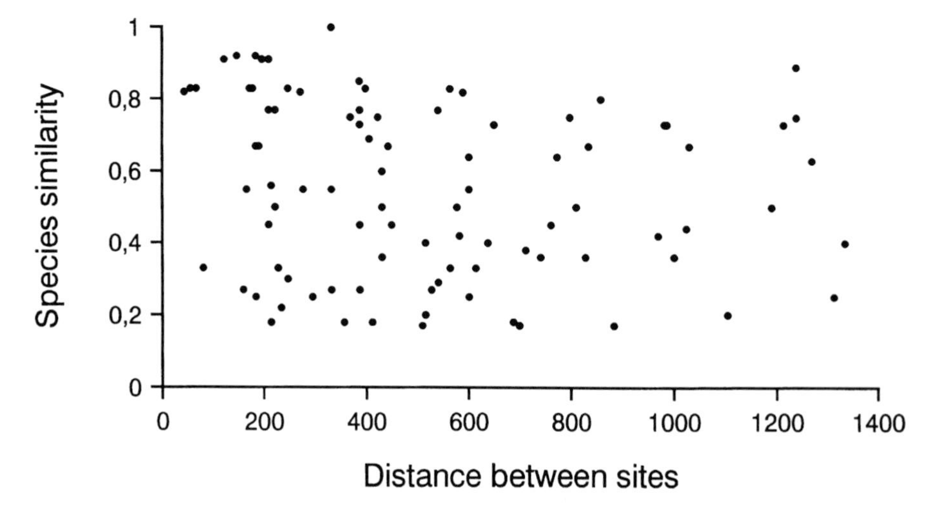

Figure 3. Jaccard's index of species similarities between sites of the eastern rain forest of Madagascar (from Ganzhorn 1998).

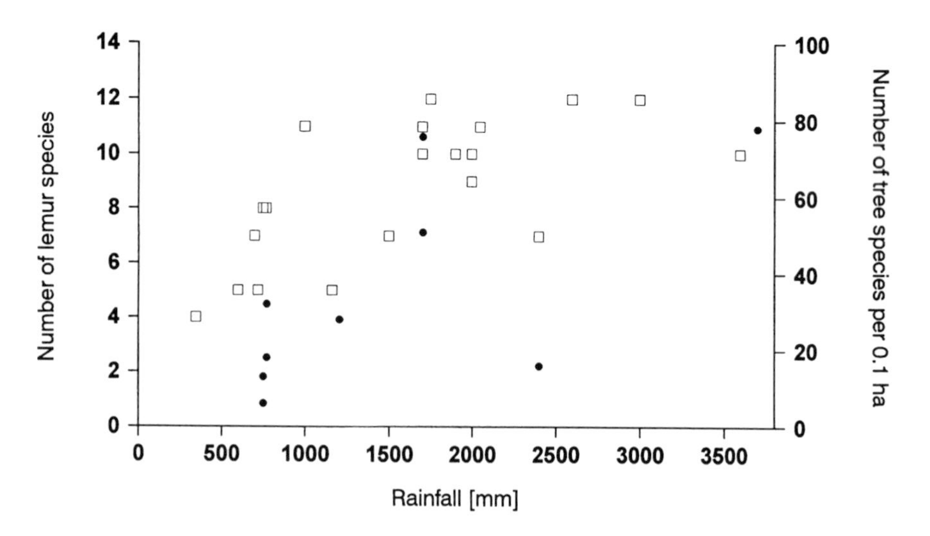

Figure 4. Number of lemur species (squares) and number of tree species (black dots) in relation to mean annual rainfall. There is a close correlation between the number of lemur species and the number of tree species in any given area (from Ganzhorn *et al.* 1997).

Effects of disturbance

Within limits of disturbance intensity and on a small geographical scale, disturbances increase forest productivity (Ganzhorn 1994, 1995) and augment habitat complexity. Lemurs reach higher species numbers and population densities in slightly disturbed areas, compared to undisturbed sites and also compared to sites which were severely modified by human action (Kruskal Wallis Analysis of Variance: Chi-square = 6.13, df = 2, $p < 0.05$; Fig. 5; Ganzhorn *et al.* 1997).

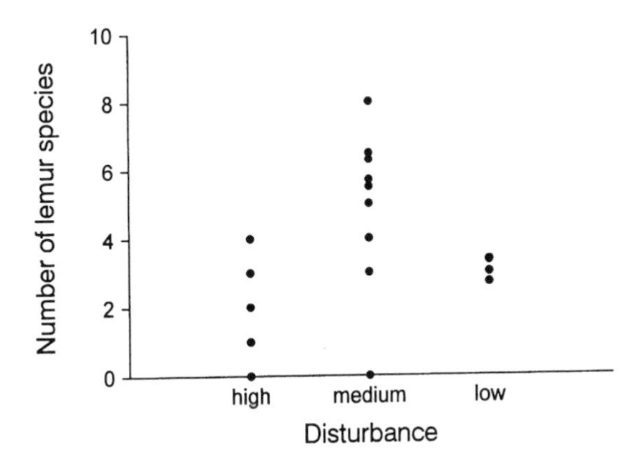

Figure 5. Number of lemur species per survey site in relation to disturbance classification of the forests at Montagne d'Ambre, Ankarana and Analamera in northern Madagascar. Sites were assigned to three different types of disturbance. High: Transformed habitats (plantations of exotic trees with or without regeneration of native trees; natural forests degraded through grazing and heavy extraction of building timber). Medium: Natural forests, disturbed by selective logging and other human activities with the effects still visible. Low: Natural forest without conspicuous signs of human disturbance, except for trails (data from Ganzhorn *et al.* 1997).

This peak curve of the number of lemur species in slightly disturbed areas may not only be related to increased primary productivity and habitat complexity due to disturbances, but rather be a consequence of selective extinction of lemur species which were unable to cope with the disturbance regime exaggerated by human interference over the last few hundred or thousand years. Even today some species, *e.g.* the small omnivorous nocturnal lemurs or frugivorous species such as *Eulemur fulvus* and even the Aye-aye (*Daubentonia madagascariensis*), seem to benefit from disturbances, especially if fruiting trees are introduced (Petter *et al.* 1977). Most living lemur species might actually belong to this type of disturbance-tolerant species. In contrast species like *Varecia variegata* are very susceptible to disturbance and especially to the removal of large mature trees (White *et al.* 1995). They might be among the last representatives of the large extinct group of lemur species which were unable to tolerate human impact.

Community composition based on functional groups

The number of different tree species, representing different potential food sources, might also determine the niche space available and how many species can partition the resources among themselves. For the Malagasy lemurs and for primates in general, different capacities for detoxification of secondary plant chemicals might be crucial for differentiation within their feeding niches (Glander 1982; Glander *et al.* 1989; Ganzhorn 1988, 1989). Many different tree species do certainly not only provide structural and phenological but also chemical diversity and thus might allow more lemur species to coexist.

The role of interspecific competition and strongly deterministic processes on a functional level is further exemplified by a test of rules possibly governing the assembly of lemur species. The rule in question here operates on a functional rather than on the taxonomic level. It specifies that: "There is a much higher probability that each species entering a community will be drawn from a different functional group... until each group is represented before the cycle repeats" (Fox 1987, p. 201). This rule implies that not species identity (*i.e.* the actual species as taxonomic unit) is relevant for community assembly, but rather that communities are composed of functional groups. These functional groups represent the units for community structure. The species within these functional groups are exchangeable.

This rule was tested with lemur communities from fourteen sites in evergreen rain forests and nine sites in dry deciduous forests of Madagascar. Lemur species were assigned to one of three different functional groups based on dietary preferences: omnivores, frugivores, folivores (classified according to Smith & Ganzhorn 1996). Favoured states are those in which the lemur species of the assemblage are distributed evenly among the three functional groups. Fig. 6 illustrates some communities in favoured and unfavoured states. In total, of the 23 sites used in this analysis, only seven belong to an unfavoured state while fifteen were assigned to favoured states as defined by Fox's rule. This deviates significantly from communities assembled at random through Monte Carlo simulations ($p < 0.001$; Ganzhorn 1997). The mismatch between random species assembly and the actual composition of lemur communities indicates very strong deterministic processes on the basis of interspecific competition not only within functional groups, but also between them. It will be one of the challenges of the future to identify and assess the precise role of these various forms of competition for community assembly in Malagasy ecosystems, and to elucidate how the functional assembly rules interact with the taxonomic constraints as illustrated by the nested subsets of community composition.

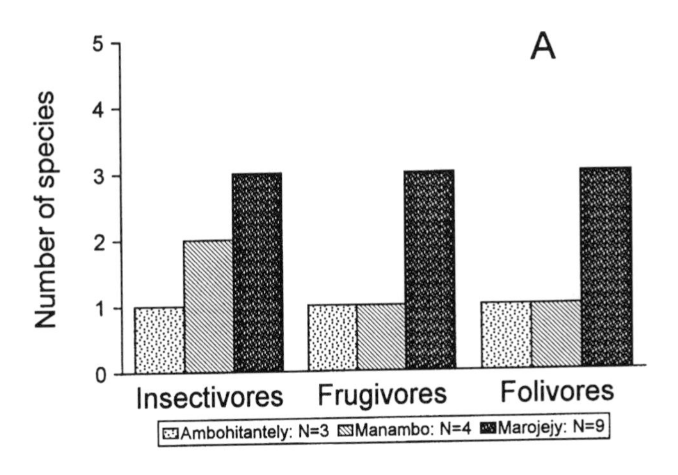

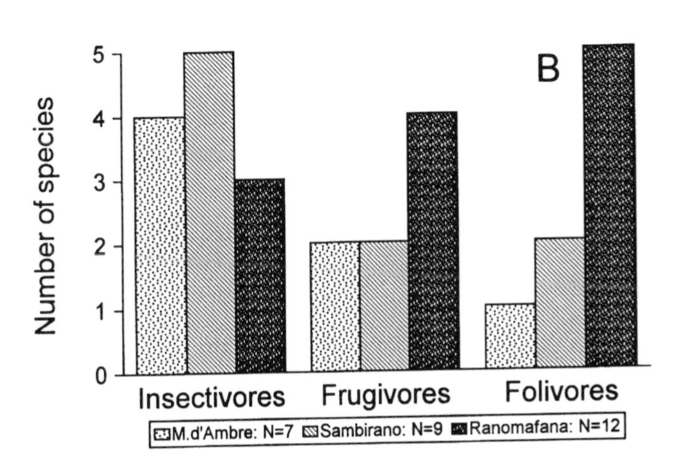

Figure 6. Examples of favoured (A) and unfavoured states (B) of lemur communities as defined by Fox's rule (Fox 1987; data from Ganzhorn 1997).

Discussion

The composition of lemur communities across Madagascar is a good example of the mutual dependencies of historical and ecological biogeography and the constraints these two concepts impose upon any interpretation of biogeographical patterns and the usefulness of species as indicators (see Myers & Giller 1988 for reviews of the underlying concepts). On the one hand, lemur species diversity shows close links to plant species diversity and thus to habitat complexity (Ganzhorn *et al.* 1997). High numbers of lemur species in any given area indicate very diverse ecosystems. If these ecosystems are modified, the number of lemur species declines, with straightforward predictions which species will decrease in numbers and drop out first (Ganzhorn

1987, 1994, 1995; O'Connor 1988; White *et al.* 1995). Thus, if an area contains many lemur species it is likely to represent an area of high biodiversity in general. This makes lemurs suitable indicators for conservation planning.

On the other hand, low numbers of lemur species or low densities do not necessarily imply that a given habitat is highly degraded. Differences in densities of an order of magnitude are common within a few kilometres of contiguous forest without obvious differences in forest structure and composition (Ganzhorn 1994). These findings can be linked in one way or another to the underlying mechanisms of ecological biogeography, even though they are not well understood in most cases. The whole scenario is then modified further by historical processes related to different vulnerability to extinction during past changes in habitat characteristics and possibly due to habitat fragmentation and isolation (Martin 1972, 1995; Tattersall 1982; Pollock 1986; Petter & Andriatsarafara 1987; Richard & Dewar 1991; Ishak *et al.* 1992; Thalmann & Rakotoarison 1994).

The present analysis illustrates how the concert of several deterministic processes operating at different temporal and spatial scales results in unpredictable composition of lemur communities. Each community seems to be subject to intense interspecific competition, determining on a functional level which type of species (species from which functional group) would be most likely to successfully invade the existing community. On the taxonomic level, species-specific characteristics might then determine which species is added to the community, based on competitive ability or tolerance of habitat characteristics. The distinction made here between functional and taxonomic aspects in the assembly of communities is somewhat arbitrary. Nested subsets of species are most likely to develop if faunas have a hierarchical organization of niche relationships (Patterson & Brown 1991). Thus, the two aspects converge on the functional level. While these considerations were based on the assembly of communities, the whole sequence could certainly be reversed and the arguments be based on differential extinction probabilities. These processes seem to be active today on the local scale of distinct lemur communities.

In contrast, large-scale deterministic colonization and extinction processes were effective in the past. Their consequences could not be compensated for in recent times. Thus, either lemurs are poor colonizers of suitable habitats, or the apparently suitable habitats are actually nothing of that kind. In either case it would be dangerous to base conservation activities for Madagascar's natural biodiversity exclusively on lemurs as indicators for habitat quality.

Acknowledgements

I would like to thank the Commission Tripartite of the Malagasy Government, the Ministère pour la Production Animale, des Eaux et Forêts and the Laboratoire de Primatologie de l'Université d'Antananarivo for their permission to work in Madagascar. Work in Madagascar has been carried out under the Accord de Collaboration between the Laboratoire de Primatologie de l'Université d'Antananarivo, the Abt. Verhaltensphysiologie of Tübingen University and the German Primate Center. My special thanks go to Mme B. Rakotosamimanana and Mme C. Ravoarinoromanga for their unflagging support of our studies in Madagascar. A. Ganzhorn, S. Goodman, A. Kratochwil, O. Langrand, S. O'Connor, B. Patterson, K. Schmidt-Koenig, U. Walbaum, and L. Wilme provided indispensable help at various stages. The projects in Madagascar were supported by the DFG, Fritz-Thyssen Stiftung, GTZ and WWF Aires Protégées (Madagascar).

References

Burney, D.A. 1987. Late Holocene vegetational change in Central Madagascar. Quaternary Research 28: 130-143.
Burney, D.A. 1993. Late Holocene environmental change in arid southwestern Madagascar. Quaternary Research 40: 98-106.
Conservation International, Ganzhorn, J.U. & Rakotosamimanana, B. unpubl. report. Geographical Priorities for Biological Conservation and Research in Madagascar. Conservation International, Washington.
Du Puy, D.J. & Moat, J. in press. A refined classification of the primary vegetation of Madagascar based on the underlying geology: using GIS to map its distribution and to assess its conservation status. In: Lourenco, W.R. (ed), Biogeography of Madagascar. ORSTOM, Paris.
Fox, B.J. 1987. Species assembly and the evolution of community structure. Evolutionary Ecology 1: 201-213.
Ganzhorn, J.U. 1987. A possible role of plantations for primate conservation in Madagascar. American Journal of Primatology 12: 205-215.
Ganzhorn, J.U. 1988. Food partitioning among Malagasy primates. Oecologia (Berlin) 75: 436-450.
Ganzhorn, J.U. 1989. Primate species separation in relation to secondary plant chemicals. Human Evolution 4: 125-132.
Ganzhorn, J.U. 1994. Lemurs as indicators for habitat change. pp. 51-56. In: Thierry, B., Anderson, J.R., Roeder, J.J. & Herrenschmidt, N. (eds), Current Primatology: Ecology and Evolution, vol. I. Université Louis Pasteur, Strasbourg.
Ganzhorn, J.U. 1995. Low level forest disturbance effects on primary production, leaf chemistry, and lemur populations. Ecology 76: 2084-2096.
Ganzhorn, J.U. in press a. Test of Fox's assembly rule for functional groups in lemur communities in Madagascar. Journal of Zoology, London.
Ganzhorn, J.U. in press b. Nested patterns of species composition and its implications for lemur biogeography in Madagascar. Folia Primatologica.
Ganzhorn, J.U., Malcomber, S., Andrianantoanina, O. & Goodman, S.M. in press. Habitat characteristics and lemur species richness in Madagascar. Biotropica.
Glander, K.E. 1982. The impact of plant secondary compounds on primate feeding behavior. Yearbook of Physical Anthropology 25: 1-18.
Glander, K.E., Wright, P.C., Seigler, D.S., Randrianasolo, V. & Randrianasolo, B. 1989. Consumption of cyanogenic bamboo by a newly discovered species of bamboo lemur. American Journal of Primatology 19: 119-124.
Ishak, B., Warter, S., Dutrillaux, B. & Rumpler, Y. 1992. Chromosomal rearrangements and speciation of sportive lemurs (*Lepilemur* species). Folia Primatologica 58: 121-130.
Kremen, C., Merenlender, A.M. & Murphy, D.D. 1994. Ecological monitoring: a vital need for integrated conservation and development programs in the tropics. Conservation Biology 8: 388-397.
Landres, P.B., Verner, J. & Thomas, J.W. 1988. Ecological use of vertebrate indicator species: a critique. Conservation Biology 2: 316-328.
MacPhee, R.D.E., Burney, D.A. & Wells, N.A. 1985. Early Holocene chronology and environment of Ampasambazimba, a Malagasy subfossil lemur site. International Journal of Primatology 6: 463-489.
Martin, R.D. 1972. Adaptive radiation and behaviour of the Malagasy lemurs. Philosophical Transaction of the Royal Society London B 264: 295-352.
Martin, R.D. 1995. Prosimians: from obscurity to extinction? pp. 535-563. In: Alterman, L., Izard, K. & Doyle, G.A. (eds), Creatures of the Dark. Plenum Press, New York.
Meyers, D.M. & Wright, P.C. 1993. Resource tracking: food availability and *Propithecus* seasonal

reproduction. pp. 179-192. In: Kappeler, P.M. & Ganzhorn, J.U. (eds), Lemur Social Systems and Their Ecological Basis. Plenum Press, New York.

Mittermeier, R.A., Rakotovao, L.H., Randrianasolo, V., Sterling, E.J. & Devitre, D. 1987. Priorités en matière de conservation des espèces à Madagascar. IUCN, Gland.

Myers, A.A. & Giller, P.S. 1988. Analytical Biogeography. Chapman & Hall, London.

Nelson, R. & Horning, N. 1993a. AVHRR-LAC estimates of forest area in Madagascar, 1990. International Journal for Remote Sensing 14: 1463-1475.

Nelson, R. & Horning, N. 1993b. Forest/non-forest classification of Madagascar from AVHRR data. International Journal for Remote Sensing 14: 1445-1446.

Nicoll, M. & Langrand, O. 1989. Madagascar: Revue de la Conservation et des Aires Protégées. WWF, Gland, Switzerland

O'Connor, S. 1988. Une revue des différences écologiques entre deux forêts galeries, une protégée et une dégradée au centre sud de Madagascar. pp. 216-227. In: Rakotovao, L., Barre, V. & Sayer, J. (eds), L'Equilibre des écosystèmes forestiers à Madagascar. Actes d'un séminaire international. IUCN, Cambridge.

Patterson, B.D. & Atmar, W. 1986. Nested subsets and the structure of insular mammalian faunas and archipelagos. Biological Journal Linnean Society 28: 65-82.

Patterson, B.D. & Brown, J.H. 1991. Regionally nested patterns of species composition of granivorous rodent assemblages. Journal of Biogeography 18: 395-402.

Petter, J.-J. & Andriatsarafara, S. 1987. Les lémuriens de l'ouest de Madagascar. pp. 71-73. In: Mittermeier, R.A., Rakotovao, L.H., Randrianasolo, V., Sterling, E.J. & Devitre, D. (eds), Priorités en Matière de Conservation des Espèces à Madagascar. IUCN, Gland.

Petter, J.J., Albignac, R. & Rumper, Y. 1977. Mammifères lémuriens (Primates prosimiens). Faune de Madagascar, Vol. No. 44. ORSTOM-CNRS, Paris.

Pollock, J.I. 1986. Towards a conservation policy for Madagascar's eastern rainforests. Primate Conservation 7: 82-86.

Putman, P.J. 1994. Community Ecology. Chapman & Hall, London.

Rakotoarison, N., Zimmermann, H. & Zimmermann, E. 1996. Hairy-eared dwarf lemur (*Allocebus trichotis*) discovered in a highland rain forest of eastern Madagascar. pp. 277-286. In: Lourenco, W. R. (ed), Biogeography of Madagascar. ORSTOM, Paris.

Remmert, H. 1985. Was geschieht im Klimax-Stadium? Naturwissenschaften 72: 505-512.

Richard, A.F. & Dewar, R.E. 1991. Lemur ecology. Annual Review Ecology and Systematics 22: 145-175.

Smith, A.P. in press. Deforestation, fragmentation, and reserve design in western Madagascar. In: Lawrence, W. & Bierregaard, O.W. (eds), Tropical Forest Remnants, Ecology, Management and Conservation of Fragmented Communities. University of Chicago Press, Chicago.

Smith, A.P. & Ganzhorn, J.U. in press. Convergence and divergence in community structure and dietary adaptation in Australian possums and gliders and Malagasy lemurs. Australian Journal of Ecology.

Sorg, J.P. & Rohner, U. in press. Climate and phenology of the dry deciduous forest of Kirindy/CFPF. In: Ganzhorn, J.U. & Sorg, J.P. (eds), Ecology and Economy of a Tropical Dry Forest in Madagascar. Primate Report, Special Volume, Göttingen.

Sterling, E.J. in press. Rapid assessment of primate species richness and density on the Masoala Peninsula, eastern Madagascar. Folia Primatologica.

Tattersall, I. 1982. The primates of Madagascar. Columbia Univ. Press, New York.

Thalmann, U. & Rakotoarison, N. 1994. Distribution of lemurs in central western Madagascar, with a regional distribution hypothesis. Folia Primatologica 63: 156-161.

White, F.J., Overdorff, D.J., Balko, E.A. & Wright, P.C. 1995. Distribution of ruffed lemurs (*Varecia variegata*) in Ranomafana National Park. Folia Primatologica 64: 124-131.

CHAPTER 3
BIODIVERSITY AND NATURE PRESERVATION

CONSERVATION OF BIODIVERSITY SCIENTIFIC STANDARDS AND PRACTICAL REALIZATION

WOLFGANG HABER

Lehrstuhl für Landschaftsökologie, TU München,
D-85350 Freising-Weihenstephan, Germany

Key words: biodiversity, nature conservation

Ecologists and particularly biocoenologists consider themselves specifically responsible for problems of biodiversity protection. But are they really always equal to this task? In order to answer this question, please try to conclusively explain to a politician in a few, easily comprehensible sentences, what biodiversity is and why it should be conserved. The political understanding for questions of biodiversity is all the more important as at the United Nations' Conference on Environment and Development at Rio de Janeiro in 1992 an international convention for the conservation of biodiversity was signed and within two years ratified by the majority of the signatories. A convention must be translated into practical measures; and for these measures laws are necessary which have to be drawn up in close collaboration between politicians and experts. Scientists, especially ecologists and biologists, are expected to make an important contribution to this process.

In this context, a word of warning is necessary. In view of the progress made in ecological research, and of the deeper insights gained into environmental problems, scientists have become more and more aware of the complexity of the relations between organisms and their environment, a complexity that occasionally causes a feeling of helplessness. We try to overcome this feeling by assigning to complexity or certain aspects thereof comprehensive "holistic" concepts, like "household of nature", natural balance, carrying capacity, environmental load, ecosystem, sustainability. For scientists, these concepts have a significant heuristic value, however, they entail considerable problems of general comprehension. The public considers them to be plausible and clear, as they seem to correspond to simple empirical facts. Someone who understands the components and mechanisms of a private household or of a municipality budget

A. Kratochwil (ed.), Biodiversity in ecosystems, 175-183

believes that he is able to understand the "household of nature", too. Consequently, a term like "household of nature" appears in environmental laws in Germany. If, however, such a law has to be implemented, severe difficulties as to its interpretation arise, which often have to be dealt with in court - of course with the help of experts, but it is judges who take the decisions. We ecologists should therefore utilize these terms more critically, especially in the face of the public, because people will otherwise assume that the terms are based on precise knowledge, whereas in reality this knowledge is only sketchy or does not exist at all.

This is also true for biodiversity. Its complexity - and the helplessness this complexity causes - are shown by the simple fact that in general linguistic usage and even within science, biodiversity is only understood as species diversity. But biodiversity involves much more, it really covers all organization levels of life on earth. In this respect, it is one of the greatest, only partly understood key problems of biology, if not of science in general. Biodiversity also influences the human and cultural scientific field, which is marked by a variety of cultures and their different aspects, as well as by the modern diversified range of views and opinions.

Biodiversity is thus a designation for the immense and therefore hardly comprehensible variety of life, its manifestations and processes. It is a characteristic of life and is actually, like some other biologically and ecologically essential qualities, an old and trivial empirical finding. Why are we again and again fascinated by identical twins? Because we are used to people being different, and the nearly perfect identity of twins impresses us so much that we immediately start to look for even the smallest distinguishing features.

Perhaps this also explains why the term biodiversity, the perfect comprehension of which is rather difficult, has found a relatively quick acceptance. That it involves a considerable political and social explosive force has either not yet been realized or is deliberately ignored. Equality, equal rights, equal values, and similar socio-political efforts, which sometimes even include animals of higher systematic rank, are hardly reconcilable with biodiversity. On the other hand, the biodiversity concept also lends itself to social or political abuse.

Within biology and ecology, the biodiversity problem is not quite a new challenge, nonetheless it provides new impulses for understanding the phenomena of life. The great biological and ecological research programmes of the last decades mainly focused on the discovery or confirmation of universal natural laws, predominantly on absorption, transformation, and processing of energy and matter. In this context, biodiversity was sometimes even considered a nuisance and was overcome or pushed aside by choosing some "standard organisms" or "domestic animals" for biological research, *i.e.* plant and animal taxa which are continually (and nearly exclusively?) used in experiments, like *Drosophila*, *Neurospora*, *Arabidopsis*, *Oenothera*, or mice, rats and rhesus monkeys, respectively.

Organisms, however, are not only determined by general natural laws, but are always the results of an evolution, *i.e.* of a process with numerous transformations and adaptive radiations, and thus each of them is unique, its evolution cannot be repeated. This fact also provides an important impulse for the preservation of organism species. In this context, a very interesting phenomenon is the contrast between continuity in the course of evolution and ramification and specialization in the corresponding time level, resulting in a discontinuous dispersal of organisms on our planet. Counteracting this differentiation is natural selection which restricts and channels speciation, but never levels it out completely. It is only modern man who has started levelling out diversity on a large scale, and must now be prevented from continuing.

Since man as consumer (in the ecological sense) is predominantly interested in the **use of resources** - but only to a far lesser extent in their **protection** - the chances for the conservation of biodiversity are not particularly high. There is certainly a connection between the functional performance of natural or man-made ecosystems and biodiversity, but it is very difficult to definitely ascertain at what point the diminution of natural diversity will really impair the functioning of an ecosystem. The hazards that might be caused by decreasing biodiversity of our economically used plants and animals can be more easily recognized. It is a well-known fact that the food supply of the growing world population is derived from a surprisingly small number of species, which, however, displayed a considerable within-species diversity as shown by a wide range of mostly local varieties or cultivars. Modern plant and animal breeding of standardized high-yielding cultivars or strains, supported by genetic engineering and cloning, is levelling out this kind of adaptive diversity, and may extremely endanger the safety of future food supply. This is an effective argument for the conservation of biodiversity, with an emphasis on wild species related to economically useful plants and animals. For the public, this argument is more convincing than ethical considerations or general scientific aspects, such as the statement that the disappearance of each species entails a loss of possible scientific knowledge.

If a scientist studies biodiversity in detail, he will first have to investigate the molecular-biological level and perhaps be surprised at the uniformity he will find there - at least at first glance. It is the uniformity of the DNA molecule, despite the apparently complicated structure of the double helix. The adenines of one of its two strands are always bound to the thymines of the other; the same is true for glycines and cytosines. Only the sequence and the relation of the two respective base pairs are different. For the physico-chemical condition of the whole molecule this is presumably of no importance, however, it seems to be essential for the expression of the genes or proteins derived from them. Thus on the one hand the combination of the base pairs adenine-thymine and glycine-cytosine is strictly determined, on the other hand there is no determination at all as to the sequence of the bases along the DNA strands, which enable life and make it at the same time diverse. The whole immense and complex biodiversity depends on this simple duality! And since the sequence of the base pairs is apparently independent of the physico-chemical state, one base pair can be replaced by another without impairing the thermodynamic stability of the molecule. It is well-known that such mutations occur frequently, on average once per cell division.

From these facts the following rule can be deduced: Biodiversity is a result of the basic characteristics of nucleic acids and of the existence of mutations (as well as of related mechanisms of exchange and modification of nucleic acids). Diversity is a fundamental quality of life; without diversity, evolution would be impossible.

From this point onwards, diversity increases on the following higher organization levels: Apart from the diversity of the protein molecules, the variety of components of the cell organelles and their own diversity play an important part. Additional influences which are not dependent on the sequence of the amino acids in the proteins take effect, *e.g.* pH-value, salt concentration, temperature, hydration, interaction of enzymes etc. As components of biodiversity, they contribute to the development of different features in the cell. At the same time, they are an expression of self-organization.

It is assumed that the mutations in the nucleotide sequence are not connected to, or influenced by, the diversity of the organismic system in which they are embedded. If this hypothesis is not refuted, it can be concluded that there are no feedbacks between system diversity and diversity-producing "mechanisms". This would mean that *active* processes for the preservation of diversity do not exist. We have much information about the effects of environmental influences on the mutation rate, however it has not yet been elucidated how heterozygosis and the diversity of higher (systematic) rank affect this rate.

At the level of the cell we recognize the first successful experiment of an individualized, autonomous organism in procaryotic and eucaryotic form. The development of a multicellular organism necessitates a further, but different diversification of cells according to their role in tissues and in the resulting organs and organ systems typical of a multicellular organism.

Thus we have now reached via unicellular (protocellular) and multicellular (metacellular) organisms the organizational level of the organism, on which biodiversity is represented by the huge variety of organisms and species, respectively. My initial criticism that biodiversity is often too much restricted to species diversity was not meant to belittle the significance of species diversity in the field of biodiversity. As individuals, organisms are the tangible "embodiments of life", and in addition the real "function holders" of the biosphere. Thus they provide concrete starting-points for biodiversity conservation measures.

As individuals, organisms are the most complex and differentiated creatures of our planet, which occur in a really fantastic variety. Their reliable identification or determination is an indispensable prerequisite for their study. This is the task of the biological disciplines taxonomy and systematics. Unfortunately, in the last decades these disciplines have been neglected to such a degree that I really fear that **species diversity** cannot be preserved because of lack of exact **knowledge of species -** or that protection is only afforded to those species which have been described and are recognizable. In its efforts to unveil the key principles of life, modern biology has lost touch with the very diversity of the "embodiments of life", and has depreciated the science dealing with it to mere registration. But taxonomy is the science of the distinguishability and differentiation of species and works out the criteria according to which the species are "sorted" and classified. It is more oriented to natural history than to general laws of nature, and it is not only its task to determine distinguishing features, but also to evaluate them. Diversity can only be determined with the help of taxonomy! In this context, it should also be remembered that taxonomic determinations are not invariable and will have to be revised if there are new research results - taxonomy, too, is a dynamic science! The incorrect determination of one species can lead to wrong conclusions about environmental changes or biocoenotic links.

In particular those ecologists who investigate populations and biotic communities must either be good taxonomists themselves, or have to be able to rely on cooperation with knowledgeable taxonomists. This claim is easily made, but can actually not be fulfilled. The species of how many ecosystems have really been completely recorded and analyzed? As a rule, we only investigate species or species groups which can be well and generally identified, and tacitly presuppose that they are sufficiently representative of the system under study. The well-known general ecosystem model contains three functional groups, of which the group of producers is taxonomically best and most generally known, even if there are still great gaps in knowledge about algae. The species composition of the group of consumers has been studied to a much lesser extent, and the group of decomposers is in many biodiversity investigations only indicated as a sum of individuals, not as individual species. There are great deficiencies in zoological and microbiological taxonomy; thus, as the "Solling project" of the IBP and similar great research projects have shown, zoologists and microbiologists are not able to draw up complete lists of species. Even if such lists were made, the species names would say very little about the species' functions in the ecosystems to which they belong, since knowledge about their biological or ecological importance is lacking.

This state of things may explain why biological diversity is often understood - exclusively or mainly - as diversity of tangible structural entities. We tend to overlook that the diversity of "ecological niches", *i.e.* of relations between organism and environment, is at least as great. Still less is known about the surprising variety of communication systems within species and populations. The diversity of behaviour expressions, caused by conditioning or imprinting phenomena, learning processes or traditions is nearly never taken into consideration, either. That is why the "Arbeitskreis für Biozönologie der GfÖ" ("Association for Biocoenology of the Society for Ecology") is a very important, valuable institution; it may be able to fill these gaps of knowledge.

It is essential to preferably investigate so-called "small" ecological systems or subsystems, like plant-insect complexes, or communities in smallest aquatical or terrestrial habitats (microhabitats). There are two approaches which ensure that biodiversity is adequately taken into account in this type of research: on the one hand the application of the concept of ecological guilds, *i.e.* of functional groups of co-existing species which use the same resources in a similar manner; on the other hand the study of food chains and food webs.

Still other, even more difficult problems arise when biodiversity on the level of ecosystems is investigated. Here, per definition, the heterogeneity of the abiotic habitat factors (for those factors, the terms "variety" or "diversity" are often avoided!) has to be included in the study. The requirement to consider biodiversity has initiated the study of new aspects in ecosystem research, a valuable addition to this field of research which had up to then emphasized - often exclusively - energy and matter transfer. Especially on the level of ecosystems it becomes clear that biodiversity based on species diversity is only one aspect of what is usually described by terms like "variety" and "richness". Future investigations must focus on interactions of species, in two different ways: 1) on the interactions between the species themselves (which would be the task of biocoenotic research); 2) on the interactions or connections of the species with their abiotic environment (which is "classical" ecology).

Only on the level of biotic communities and ecosystems, changes in species diversity can be ascertained and interpreted. Many ecosystems show impoverishments by decrease in or disappearance of constituting species; but in some cases species diversity can also increase. A short-term increase in species diversity may be caused by immigration, a long-term raise by species formation, which however will actually only be ascertained with species of short generation turnover. The importance of scale is still underestimated in this context. By choosing different scales it could be "proven" that ecosystems of temperate or Mediterranean zones are richer in species than those of the humid tropics! The same is true for the time dimension, the much and controversially discussed "stability" of biotic communities and ecosystems, respectively. The trivial fact must not be forgotten that organisms (level of species) differentiate **themselves** from their environment, whereas ecosystems are largely delimited by the investigator, according to the formulation of a particular question. Ecosystem delimitations "are less found than made"!

This problem is of great significance for the interpretation of diversity changes. It is by no means always possible to exactly indicate or assess how many changes in diversity are due to human influence and what part "natural" factors play - which, however, may be indirectly affected by humans. The relation to the overall biogeographical distribution of a species, which is often not taken into consideration in local or country-based studies, is also important. Among the plant species which became extinct in the former Federal Republic of Germany (until 1990) there was only one which solely occurred here and was thus really eradicated. Flora and fauna of the country have of course suffered considerable losses, and in small areas (viz. the scale problem) these losses can amount to nearly 100 %. Responsible for this are predominantly "interventions on the spot", which are not always due to modern land-use methods, but may also be caused by the traditional techniques of agriculture and forestry.

What relation is there between biodiversity and complexity? Complexity is defined as something very intricate or complicated. But living systems are also complex in another way: they contain a wealth of information - such as a play by Shakespeare is more "complex" than a children's book - and they are subject to highly complicated dynamics. This touches on fundamental problems of scientific theory, among others the reductionism-holism controversy. The reductionistic approach necessitates the investigation of a system on the level of its single components in order to be able to understand it. Someone who wants to gain detailed insights into the behaviour of species has to study their physiology and genetics; likewise, to comprehend the dynamics of biotic communities, the behaviour of individual species must be explored etc. The holistic approach sees in systems more than just the sums of their components and aims at newly emerging qualities. In this connection, the concept of the hierarchical organization of living systems should again be emphasized.

The above controversy seems to me superficial, because the holistic conception by no means excludes a strictly analytical approach, a splitting up into essential components, in order to elucidate the characteristics of systems - and thus their biodiversity. However, you may choose if you prefer to divide up the whole system ("holon") into its single parts or first into its lower levels of organization. It is misleading, too, to understand biological systems exclusively as the result of the combined action of their components; lower levels in the hierarchy are rather determined by phenomena of higher levels. A cell in a multicellular organism is shielded from its abiotic environment, but this organism is as a whole subjected to the conditions of its surroundings, and its response to them determines the behaviour of every cell - but in a different manner. A higher degree of complexity enables some organisms to "disregard" certain environmental conditions, but excludes the utilization of other environmental assets. For example, only multicellular organisms are able to move by themselves or to fly, but only procaryotes can live in hot springs or convert nitrogen into ammonia. A single procaryote capable of photosynthesis can probably exist on its own infinitely by fixing nitrogen and oxygen from the air and absorbing other nutrients from its surroundings, *e.g.* from rain water. Higher plants can fix a considerably higher amount of carbon and absorb other nutrients more quickly, and they thus produce significantly more organic substances - but, since they cannot fix nitrogen, they cannot forever exist by themselves. They clearly depend on co-existence with procaryotes, which decompose dead organic matter and thus release the nutrients contained therein, including nitrate.

These examples show that biological complexity arises from life processes which take place on different spatial and temporal scales. The fact that these different scales have often been disregarded in the past has led to much confusion. This is particularly true for the investigation of populations, biotic communities, and ecosystems.

There is much speculation today on the deeper significance of biodiversity, however, no clear evidence can be provided. If all enzymes encoded in the DNA of one cell served the same function, there would be no differences in metabolism, and the development of new forms and new levels of complexity in the organization of the cell would not be possible. If all cells had the same characteristics, there would be no differentiation on the level of tissue; likewise, diversity of species is obviously a prerequisite for the structure of biotic communities. But which degree of variety, or which type of combination of species is actually necessary for the smooth functioning of communities and ecosystems? As yet we are not able to answer this question, something environmental policy-makers blame us for. This touches on the old controversy whether a biotic community consists of species which have gathered at a given place by pure chance, or if their presence is more than merely accidental. The same is true for the formation of patterns in communities or ecosystems, respectively; the existence of these patterns may not be doubted, but it has not yet been clarified how they develop - are they the result of interactions between species and their abiotic environment, or have they arisen accidentally, too?

The question if complex systems are more stable than simple ones is also loaded with controversy. Already more than 20 years ago, Robert May showed at model systems that complex systems are not necessarily more stable than simple ones - but did he choose the right model systems? Functions have to be considered, too. Animals are functionally more diverse than plants, and microorganisms, in turn, are more diverse than plants and animals. Animals and microorganisms perform more, and more varied, physiological functions, and use a higher amount of different resources than plants. Among animals, on the other hand, there is a high degree of functional similarity, as demonstrated by the concept of the "ecological guilds". This again causes redundancy - and thus a "dispensable" part of diversity. People who oppose the preservation of all species here find their arguments. But on closer examination of the statistical distribution of species functions it can be shown that most functions are fulfilled frequently, others however seldom or just once. Such rare functions are often only found in communities rich in species, like triangle, harp, or kettle drum are only played in large symphony orchestras and not in small bands. Likewise, the so-called key functions and their carriers can be identified. In many cases, ecosystems display degrees of biodiversity which are a lot higher than necessary, at least for an efficient functioning of their trophic system. In a large-scale analysis, biodiversity has no importance for the balance of carbon, other nutrients, and water. If small-scale analyses are performed, the opposite may be the case!

Thus we have already reached the so-called highest level of biodiversity, *i.e.* the spatial diversity of ecosystems or the diversity within biomes. Questions concerning the complexity of a landscape and of its physiognomy arise in this context, too; however, the investigation of this complexity is often strongly influenced by certain subjective values, because landscape is an ambiguous term. In landscape ecology, both dispersal barriers and connections (which are mainly founded on species, but also caused by currents of air and running water) play an important part; in English, this is frequently called "connectivity".

On all these levels above the organism level it has to be borne in mind that scientific handling of biodiversity is based on biological taxonomy. In this connection, I would like to deplore once more both the neglect and the shortcomings of this discipline, and the resulting problems for the maintenance of biodiversity. The general knowledge of species is declining, and many population biologists and biocoenologists have gathered the necessary knowledge of species within their special fields by private learning. Even then this knowledge is still limited, however, the advantage is that it includes also knowledge about different life-forms, environmental responses, and behaviour, something modern taxonomists do not necessarily possess, since they are first of all experts for differential characteristics of taxa. They use modern laboratory techniques, like electrophoresis, and as a consequence lose contact with field work and the "real world". Without the assistance of the "hobby taxonomy" of interested laymen, biodiversity research would not be possible.

However, this only characterizes the relatively favourable situation in the highly developed industrial countries, which are centres of modern scientific research and education. In the developing countries of the so-called Third World, where biodiversity still abounds in large areas, there are far less research facilities for its investigation, and the scientific competence is often inadequate. The research work performed there is mainly done by scientists' teams from industrial countries. Studies with surprising, even pioneering results had been carried out with expensive technical equipment in the tops of trees in tropical rain forests, in the course of which a vast amount of new species and species combinations had been discovered. Unfortunately, most of this research work in the tropics lacks continuity which would be essential because of the rapid and mostly negative environmental changes occurring there. As a rule, there are only research programmes with a duration of 2 to 5 years. After their expiry, the research infrastructure and the accumulated local experiences often get lost.

Finally, it is a deplorable fact that the economic and political stability which is so very important for field work or extensive outdoor research, is increasingly shaken or going to disappear altogether. Again and again it has been asserted that since 1970 the awareness of environmental problems and issues has grown and is indeed influencing human actions. However, this turns out to be rather the illusion of a comparatively small, idealistic, and educated group of people who expect others to think and act like they do. The experiences only of the last decade have shown that the behaviour of people and nations is driven by quite different priorities; the great Environment Conference of 1992 has obviously not been able to really change attitudes. Where people fight each other and become entangled in bloody combats, the environment and thus its biodiversity are seriously damaged, too; however, this is hardly ever mentioned. Who thinks, in view of the horrible slaughter which has recently taken place in Rwanda, of the fact that this civil war may also seal the fate of the mountain gorilla population living at the border between Rwanda and Zaire, as well as that of many other endemic species? Efforts to protect and preserve these unique examples of biodiversity, for which the industrial countries have spent millions, are thus ruined forever. This is an especially painful example for the fact that certain political, economic, and social interests, which allegedly serve to preserve or further promote civilization, are always enforced at the expense of diversity and complexity in nature.

References

Parts of the article are based on a special publication of the "Deutsche Forschungsgemeinschaft" entitled "Biodiversität". Moreover the following publications were analyzed:

Solbrig, O.T. 1994. Biodiversität. Wissenschaftliche Fragen und Vorschläge für die internationale Forschung. Bonn, Deutsche UNESCO-Kommission (MAB).

Solbrig, O.T., van Emden, H.M. & van Oordt, P.G.W.J. 1992. Biodiversity and Global Change. Paris, International Union of Biological Sciences (IUBS), Monograph No. 8.

UNEP 1993. Global Biodiversity. UNEP/GEMS Environment Library No. 11. Nairobi: UNEP.

World Resources Institute (WRI) *et al.* 1992. Global Biodiversity Strategy. Washington, D.C., USA.

After completion of the manuscript, I came across another important publication on biodiversity conservation problems which I recommend reading:

Raustiala, K. & Victor, D.G. 1996. Biodiversity since Rio: The Future of the Convention on Biological Diversity. Environment (Washington) 38: 17-20; 37-43.

ELEMENTS OF BIODIVERSITY IN TODAY'S NATURE CONSERVATION DISCUSSION - FROM A GEOBOTANICAL VIEWPOINT

H. HAEUPLER

Spezielle Botanik, Ruhr-Universität Bochum, Universitätsstr. 150, 44801 Bochum, Germany

Keywords: biodiversity, nature conservation

Introduction

One of the today most frequently used terms in discussions about nature conservancy is "biodiversity". A word which "trips off the lips of many a politician nowadays, but how many of them truly understand what it means?" (Akeroyd 1996). That is without any doubt true, not only for politicians but to the same extent for many scientists. The growth of literature about this phenomenon, voluminous books included, seems inversely proportional to the growth of knowledge of what biodiversity really is, or better: is at a given distinct place on earth.

The discussion in the nineties[1)] began with Wilson's book simply titled "Biodiversity" (1988), which was translated into German in 1988 with the catchpenny title: "Ende der biologischen Vielfalt? Der Verlust der Arten, Gene und Lebensräume und die Chancen für die Umkehr" ("The end of biological diversity? Loss of species, genes and habitats, and chances for a reversal of the development"). In a certain sense this German title indicated some of the main problems behind this simple term. Article 2 of the "Convention on Biological Diversity" (IUCN, Rio de Janeiro 1992) defines it as "the variability among living organisms from all sources including, inter alia, terrestrial, marine and other aquatic ecosystems and the ecological complexes of which they are part; this includes diversity within species, between species and of ecosystems". This definition implies the variability of life in all forms and combinations and on all levels. After Akeroyd this does not mean "only the sum of all ecosystems, species and genetic material. Rather it is the variability between and among them". Biodiversity is therefore an "attribute of life", analogous to productivity, mobility, reproductivity or similar aspects. In the following Courrier (1992) drew up a "Global biodiversity strategy" with 85 more or less concrete suggestions for actions, similar theoretical proposals were made by McNeely *et al.* (1990) and Solbrig (1991).

The ICBP (1992) gave examples for "putting biodiversity on the map", but only for some birds inhabiting an area of less than 50,000 km². Kheong & Win (1992) edited

1) The term "diversity" as a measure of species richness was used for the first time in 1949 by Simpson (not by Fischer *et al.* 1993, as often cited). The Brookhaven Symposium (Woodwell & Smith 1969) made it a magic term, followed by a lot of papers (see the history in Haeupler 1982, 1993).

A. Kratochwil (ed.), Biodiversity in ecosystems, 185-197

with "In harmony with nature" the proceedings of the "International Conference on Conservation of Tropical Biodiversity" which primarily deals with Malaysian ecosystems. Bellamy described in this volume diversity as the "spice of life". The Malaysian colleagues give considerably concrete details of special aspects of diversity. Rickleffs & Schluter (1993) edited a selection of manuscripts about "historical and geographical perspectives of species diversity in ecological communities". Doubtlessly it contains a lot of interesting and remarkable papers. In "Biodiversity and landscapes. A paradox of humanity" (Kim & Weaver 1994) it is pointed out that humanity depends on biodiversity and landscape systems for its survival, yet, at the same time every use of living resources places the existence of these natural systems at risk. Huston (1994) reviewed patterns and theories of biodiversity and published a synthesis in his "Biological diversity, the coexistence of species on changing landscapes". "Widening perspectives on biodiversity" by Krattinger *et al.* (1994) will be dealt with later on.

The here mentioned books are only the tip of an iceberg. They contain hundreds and thousands of pages on glossy paper, there are laws, there are strategies to save nature, there are conventions on this topic, there are numerous theoretical projects on biodiversity, on the whole there are a lot of "good politics but poor science" in the sense of Akeroyd (1996).

One of the most striking consequences of the concept of "diversity" which has been discussed for nearly 30 years is that such widely diverging aspects as diversity of ecosystems, species (plants and/or animals), genes and (last but not least) molecules are unified in only *one* scientific term: "diversity" (or today "biodiversity"). The only way to define such a complex network of interactions as "biodiversity" is to divide it into its essential elements. For me one of the most convincing examples of defining complex structures in scientific literature is the synoptical splitting of the term "neophyte" by Schroeder (1969). This, too, was a problem discussed for decades without any chance of solution in only *one* single scientific term. I myself had suggested in 1972 that "diversity" should be separated strictly into diversity of species, diversity of structure, diversity of dominance, diversity of ecological valence etc.

"Global biodiversity; status of the earth's living resources" by Groombridge (1992) is the most comprehensive global survey of what is known about this topic world-wide. Let it be the base for some of the following remarks. I want to show how uncertain this is.

What are the elements of biodiversity?

It begins with the as simple as difficult question "what is a species?". In the mentioned IUCN book it is answered by the conclusion that "for the present we have to manage with a very imperfect and inconsistent system of classification, even at the... level of species. In practice we have not advanced much beyond the position outlined long ago, that a species is what a competent systematist says it is, formulated 1926 by Regan." An indeed little encouraging remark at the beginning. And "species" are one of the most fundamental parameters biodiversity is defined by!

Merxmüller, a reputed German systematist had pointed out in 1966 that about 200 years of intensive systematical and taxonomical work were required before having a nearly complete inventory of the flora of the earth. After an estimation in Groombridge (1992) plants represent only 2.4 % of possibly existing species of the global total (Fig. 1)! Confronted with these impressive relations, documented on 585 pages in the cited IUCN balance, one is overcome by a deep frustration or rather helplessness. But at a second glance one is on the other hand astonished at the courage or rather the impudence with which parameters only based on estimations are juggled with. These estimations are often founded on hair-raising forecasts, *e.g.* the often repeated estimation of Erwin (cited in May 1992). From the species number of *beetles* on only one in detail studied tree specimen in Costa Rica, the total species number of *insects* in the world is extrapolated by forecasts. But there is neither any arithmetical nor geometrical or otherwise provided correlation between the species number on one tree and that on another one, especially not between those in different climatic zones.

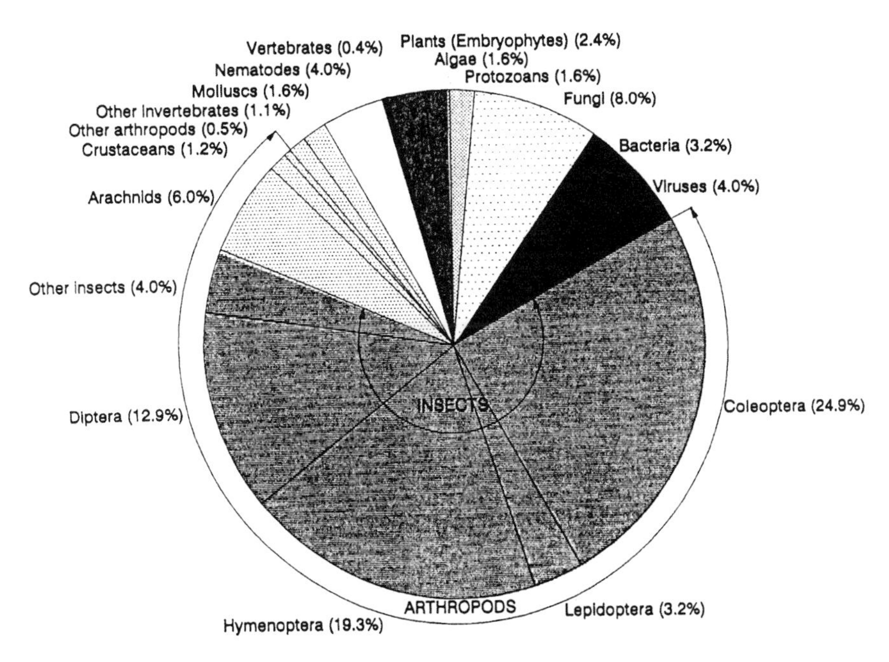

Figure 1. Major groups of organisms: possibly existing species as proportions of the global total. (Possible proportions of major groups of organisms based on conservative estimates providing a total of approximately 12.5 million species for all groups. All groups considered likely to contain more than 100,000 species are separately given in the pie-chart, along with invertebrates for comparison. From Groombridge 1992.)

Sometimes scientists are forced to make such statements. In practice we are often asked to comment on problems of our time at once, without enough time to verify the data, *e.g.* those of biodiversity. If we are fair, we must admit that we do not know very much about this subject. We therefore have no choice but to work with estimations. But it becomes problematical when scientists employ such estimated parameters in further, more detailed statistical analyses, ecological modelling and other generalized rules. Especially in Anglo-American publications you will find a lot of such inadmissible generalizations of statistically insufficient samples (see many examples cited in Haeupler 1982, in Begon *et al.* 1986 [*e.g.* Fig. 20.1] or Fig. 5.3 in Groombridge 1992). From only eight data a linear function was derived and postulated as a general rule, *e.g.* as species-area relationship. But the more data you have, the more you will find a sigmoidal swarm and no linear function (Connor & McCoy 1979; see also Williams 1964; Cailleux 1969 and Haeupler in press).

In order to gain more information, further systematical research is necessary, modern floras have to be drawn up for every part of the world. But the species level is only one segment in the scope of biodiversity. Other elements of diversity have to be considered.

Without any reference to what is diverse, the term has no substance. (Bio)diversity does not exist as *the* probability per se (it exists only *from* something) or as *the* information (it exists only *about* something). We have to distinguish between species diversity, structural diversity, habitat diversity etc. The α– to γ–diversity by Whittaker (1972) is the first step, but in view of the complexity of nature only a very, very small step. If you compare the whole literature about diversity, you will always find only four aspects accepted by nearly all contributors: genetic diversity, species diversity, habitat diversity and landscape diversity. But if the levels of organisation of life on earth are investigated (Fig. 2), a lot more elements can be found worth being considered under diversity aspects. Apart from the levels of diversity in biotic systems, similar levels of diversity in cultural and human behaviour, historical and political relationships could be set up.

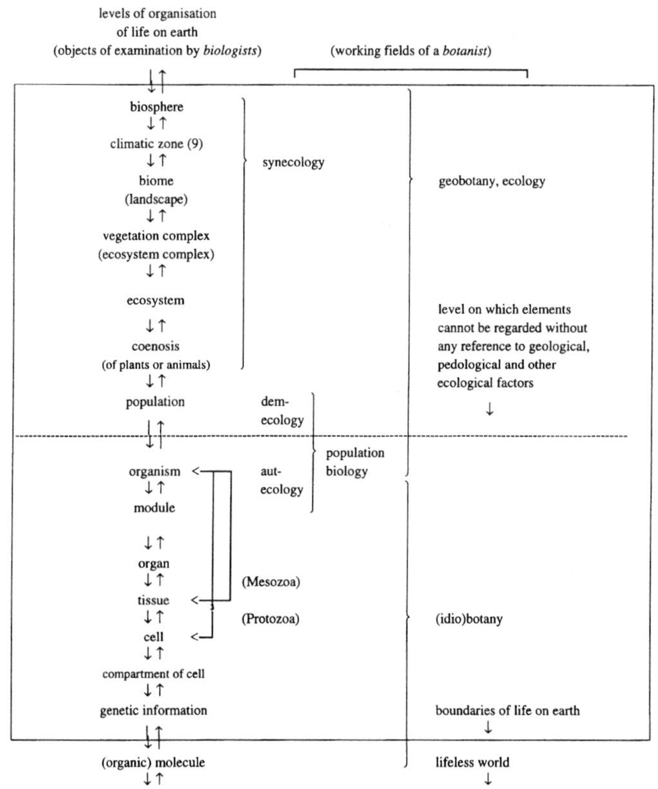

Figure 2. Levels (operational units) of organisation of life on earth (extensively completed and differentiated after many authors).

The most fundamental gap in every biodiversity discussion up to now is the lack of knowledge on the population level. On this level most phenomena of competition and survival took place, which determined the structure of the following levels in this hierarchy of operational units. But what do we know about these population levels? It's something like nothing, a drop in the bucket.

Fortunately a new generation of literature seems to develop, a new view of biodiversity taking into account this population level (*e.g.* Schulze & Mooney 1994; Duelli 1995; Matthies *et al.* 1995; Frankel *et al.* 1995; Heywood & Watson 1995; Akeroyd 1996; Gaston 1996 and Vitousek *et al.* 1996). Magurran (1988) and recently Hawksworth (1995) published valuable books on methods for measurement and estimation of biodiversity.

What to do?

In view of these serious gaps in our knowledge of basic biodiversity data, fully accepted by the above new literature, one is all the more irritated to hear about "Widening perspectives on biodiversity" (Krattinger *et al.* 1994). This is a voluminous book with papers of the "Global Biodiversity Forum Meeting" in Gland, Switzerland, in October 1993. For a geobotanist it is an irritating and alarming book. Thailand's experiences in the field of biodiversity documented by Sriwatanapongse (1994) are representative of how the matter is seen by the IUCN. The "tremendous wealth of plant and animal species" has often been cited for Thailand and considered in several acts for nature conservation, too, but the paper focused on the genetic resources of maize (!) and rice. Biotechnology uses the vehicle "biodiversity" to get more publicity! The myriads of field workers in the world who are confronted with diversity every day in their local work are not involved. Their struggle for conservation of real specific diversity is not reviewed in these books, is not funded by the agencies of the UNESCO, the IUCN and the European Commission. If that is the conservation policy of the 21st century, as so often pointed out in this volume, it is a rather bleak and hopeless situation for real nature conservancy. This way of administrating biodiversity by agencies and relevant authorities, both without any biological expertise, by conferences about programs, acts and conventions is no solution while nature is rapidly more and more vanishing. 473 pages of papers from directors, co-directors, executive directors, officers, members of environmental committees are presented. Where are the biologists? A real "widening of biodiversity"? I can only repeat the words of Akeroyd (1996): "good politics, *poor science*". The need for measurement and evaluation of biodiversity reclaimed in this volume, *e.g.* by Senanayake (1994) has not made any progress beyond the α– to γ–diversity of Whittaker (1972). Biodiversity is only understood as "the measure of the variation in genes, species and ecosystems" (McNeely 1994, Chief Biodiversity Officer of the IUCN). There is no progress in thought. New goals, strategic lines are postulated, biotechnology is seen as "deus ex machina", but no perspectives for real life can be detected.

But what can we do? We need basic research, the precise count of diversity on every level (see Fig. 1) at as many as possible concrete stands and with statistically sufficient sample sizes to evaluate, to duplicate all data on biodiversity. Reliable checklists of all types of organisms are the first step to come to secured counts. A solid analysis of population processes is another one. All well-formulated strategies, all conventions will fail if they cannot be based on reliable data from concrete stands. Here we must begin to investigate, to invest money. The results will not be spectacular, but without them, no politician will succeed in translating the Convention of Rio de Janeiro (1992) into concrete measures.

Signor (1990) reviewed the taxonomic diversification in the earth's history. Many physical and biological processes influenced diversity through time. Larger and minor extinction events had caused a decrease in diversity from time to time. But all in all, in spite of a very controversial discussion, the general pattern that species diversity and complexity of other biosphere elements had been continuously increasing during geological periods is agreed upon the majority of scientists. But today humanity is at the starting point of an interruption of this biodiversity increase of perhaps never seen intensity.

We cannot maintain existing biodiversity by protecting sites alone. To preserve biodiversity the interactions among species should be preserved. "Pattern of species diversity can often only be understood if the underlaying population dynamic mechanisms and pattern of genetic diversity are also studied" (Matthies *et al.* 1995). But Schulze & Mooney (1994) pointed out: "our knowledge is weak in important details, however", *e.g.* "what is the role of an individual species in ecosystem function?".

The concrete study of biotic interactions between individuals shows "how slow gradual changes in environmental conditions may have serious consequences for biological diversity", like the increase in atmospheric CO_2 (Matthies *et al.* 1995). The authors cited a study of Körner who had shown that the decreasing populations of *Gentianella germanica* are not influenced by the CO_2 itself, but by the altered competition: *Gentianella* could no longer successfully compete with the other species of the plant community supported by CO_2.

How to evaluate biodiversity?

As reliable indicator for biodiversity, either the simple number of elements which compose a level of organisation of life on earth (see Fig. 1, *e.g.* "species") or the interactions between the elements can be used. We have only just begun to understand these interactions which may maintain ecosystem functions (see Schulze & Mooney 1994; Matthies *et al.* 1995). Today still nobody can answer the question: what happens if a plant (or animal) species becomes extinct?

If we consider only the simple number of elements it need not to be expressed in mathematical or statistical formulas. It could simply be counted (or estimated). Such a census can be put on maps (Fig. 3), the same method was used by the ICBP in 1992.

A way to indirectly evaluate interactions of elements is calculating the diversity by measures of information theories, first of all the so-called Shannon-Weaver-index H' (see Haeupler 1972, 1995; Magurran 1988). Haeupler (1982) had shown that this index depends on n = the number of elements of the sample and therefore is not comparable with any other H'-value if it is not related to the so-called evenness $E = H'/H_{max}$ $100 = H'/\log n \times 100$.

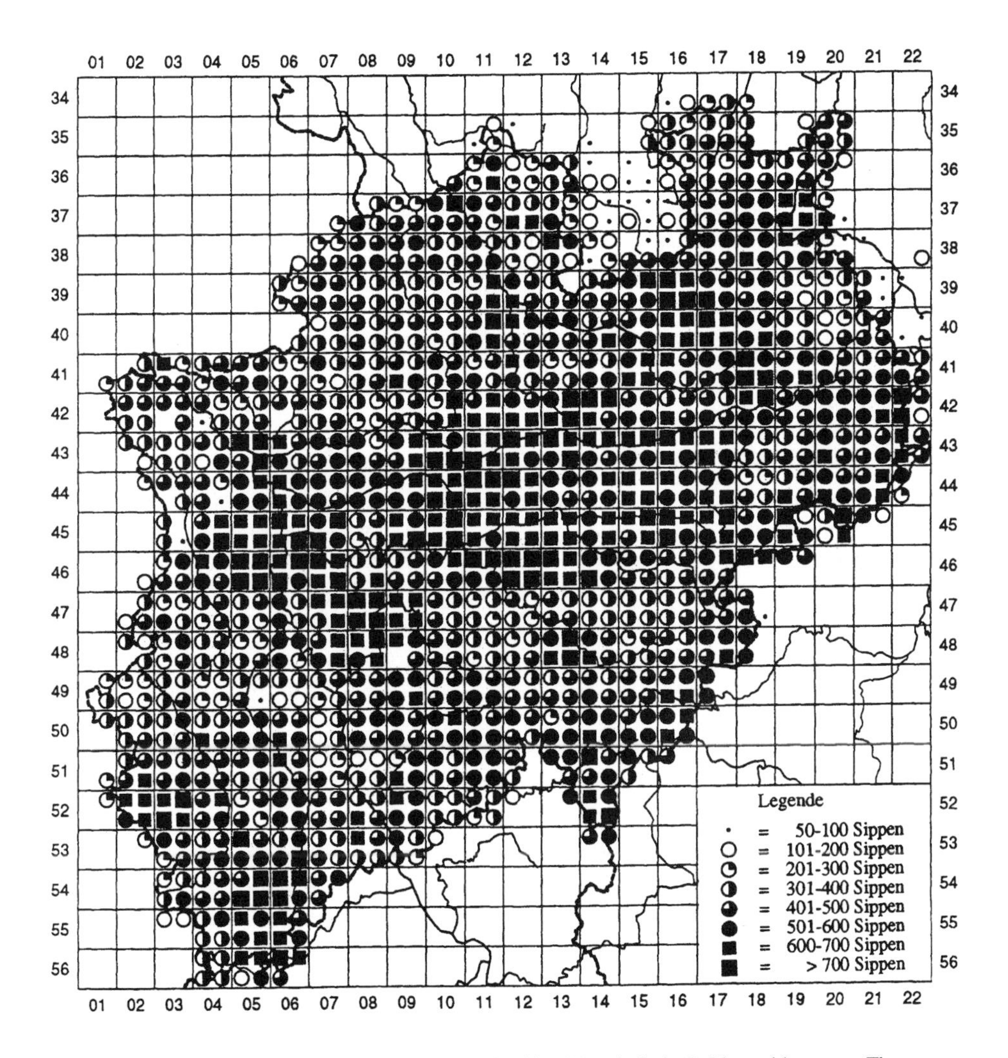

Figure 3. The number of higher plant species in North Rhine-Westphalia in 5x5 km grid squares. The centres of species diversity lie either in the industrial region of the Ruhrgebiet (centre of the map) or in the limestone mountains (Eifel in the southwest, Teutoburger Wald in the northeast of the map), that means in regions with very unequal quality. Original.

The parameter E (not H'!) is suitable for an evaluation of changes, perhaps in the structural regard of ecosystems. So in vegetation the composition of species seems nearly constant in the course of the years, but the dominance of each species can considerably change from year to year. Haeupler (1982) had shown many examples of how sensitively the parameter E can indicate stress situations in vegetation. Fig. 4 shows a wet heath with and without plucking. Fig. 5 shows the intensity of stress, which can destroy the typical structure of a given vegetation unit. The "diversity-begat-stability"-dogma by McArthur (1955) has proved to be a real non-concept. The dogma is ascribed to Elton (1958), but this author did not use the term "diversity", he spoke of "variety, richness and diversification of landscape", a much more complex statement! The equation with the Shannon-concept cannot be realized in spite of hundreds of papers in which quite the reverse is stated. The dogma is rejected *e.g.* by Begon *et al.* (1986). There is no reason to maintain this non-concept, although many ecologists and administrative biodiversity officers still do today.

On the level of species (bio)diversity it has to be distinguished between 1) variability, that means the number of elements involved in constructing an ecosystem, and 2) abundance, that means the reliable population, *e.g.* as coverage of foliage, as number of individuals of a species or as the rates of productivity.

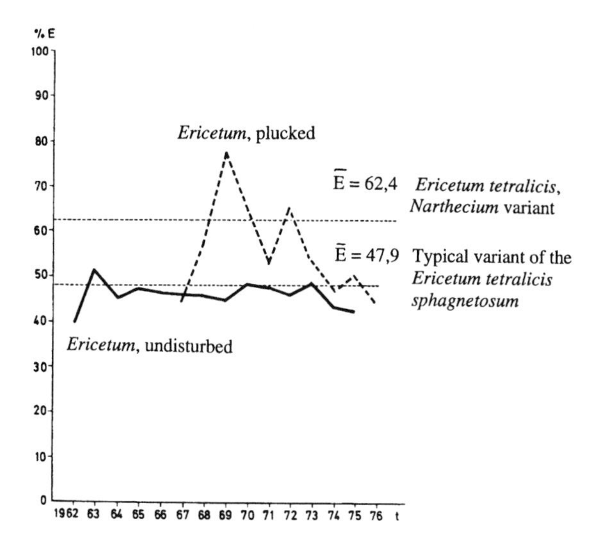

Figure 4. Having plucked a wet heath (*Ericetum gracilis*) the evenness changed dramatically. After some years of succession the evenness came up to the same values as on the undisturbed site (after Haeupler 1982).

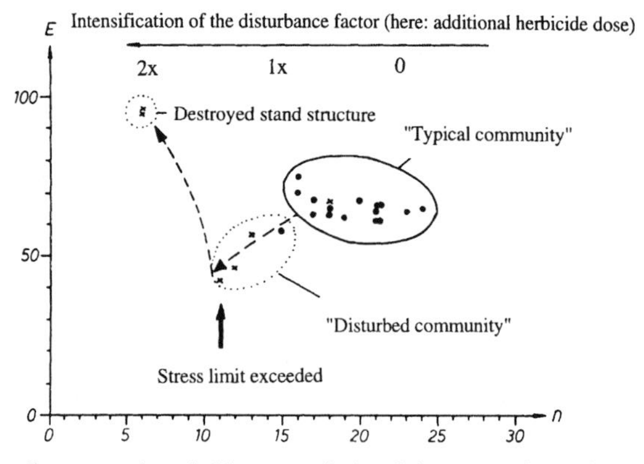

Figure 5. Diversity diagram and probable stress limit of the vegetation unit Euphorbio-Melandrietum, made visible by changing the evenness. Between the "typical community" and the minimum values of E, the transitions, also qualitatively, are fluid. For the time being, the synsystematic unit is preserved. Only when the double dose of herbicide is used, a distinct change can be observed, with extreme values of E nearing 100 (n = 6), accompanied by a complete dissolution of the stand structure and quality. (After Haeupler 1982).

How to record phytodiversity in the field and how to put it on a map

The most simple way to record phytodiversity is to count all plant species in a given area. How uncertain such numbers of plant species still are can be seen in Groombridge (1992), Tables 4.3, 8.2 and 8.3. The discrepancies between described and estimated (Table 4.3), between cited and real numbers (Tables 8.2, 8.3) are extraordinarily high. So even up to now an unquestionable number of vascular plant species for Germany, the flora of which is accepted as well as explored, does not exist. In the German check-list (Wisskirchen & Haeupler 1997), about 3,800 taxa are listed, much more than ever estimated or wrongly "counted" in common field floras.

Up to now reliable numbers of species per area can only be derived from mapping projects like Perring & Walters (1962) or Haeupler & Schoenfelder (1989) and many successors of these pilot mapping schemes in Central Europe. In Fig. 3 the most current census from North Rhine-Westphalia is given. The centres of species diversity lie either in the industrial regions of the Ruhrgebiet or in the limestone mountains (Eifel, Teutoburger Wald), that means in regions with very unequal quality. This demonstrates clearly the ambiguity of such a census of simple species numbers per area.

Attempts were made to estimate the species diversity of vascular plants on a global scale. The first map of such an estimation was published by Malyschew (1975), the most current is that of Barthlott *et al.* (in press). It is not the place here to discuss the fundamental differences between the two maps. In a general sense there are some essential similarities. But the map of 1996 is much more detailed and the hot spots of species diversity marked in red (that means more than 3,000 vascular plants per 10,000 km^2) are topographically more precisely delimited than in the first map. These hot spots are *e.g.* the Mesoamerican Cordilleras, the Central Andes, Cuba, Guaiana, south-eastern Brazil, the high volcanic mountains in eastern Africa, the southern tip of Africa, Madagascar, the south-eastern Himalaya and Yunnan, south-eastern Asia and the whole Indonesian Archipelago, and a small area in north-eastern Australia (Fig. 6).

Fig. 7 shows the attempt to describe the hot spots of plant-species diversity in Germany. For the above-mentioned reasons a map depicting merely the species number per area is too ambiguous. Therefore I have taken as a basis the number of endangered and/or rare vascular plants given in the Red Data Books. In this preliminary map[2)] some regions in Germany are striking because of an accumulation of grid squares (approximately 11x11 km) with more than 72 endangered and rare species. From north to south these are: the river Elbe valley, the central Harz Mountains, the east German Dry Region in the rain shadow of the Harz Mountains, the river Rhine valley, the large bend of Jurassic mountains called "Alb" and the foreland of the Alps. These regions are most problematical for nature conservancy; here most conflicts between conservancy management and economically oriented land using interests arise. In the course of an analysis of the recent mapping project in Germany we will prepare more precise maps. This example is only the first step.

At the end I will give some examples of the very few really existing empirically founded biodiversity studies in which phytodiversity is recorded at selected real sites in the field.

Seine (1996) has studied the biodiversity of monolithic inselbergs in Zimbabwe. He comes to the conclusion that "a combination of deterministic and stochastic mechanisms seems to control the diversity of inselberg vegetation".

2) I thank Rudi May very much for preparing this map!

In the same scientific program involving a study of the mechanisms of tropical diversity for its maintenance, Ibisch (1996) investigated the neotropical epiphytic diversity in Bolivia. A relatively high succession dynamics in epiphytes is documented. Only within given climatic conditions the epiphyte diversity is predictable, *e.g.* species number per area. The regional mosaic of habitat quality determines the degree of phytodiversity (and I think, too, of zoodiversity). The highest phytodiversity of epiphytes was found in the montane rain forest but "the epiphytes, however, here contribute no more than 20% of the species... The maximal diversity is observed where the most favorable resource combination is given." Specific biotic interactions between epiphytes and their host plants (porophytes) do not contribute to the diversity in "an exceptional way".

I refer to this paper of Ibisch (1996) in so much detail because it seems to me a very good example of how biodiversity should be: studies at concrete stands in the field and with an adequately large sample size (in this case 6,800 collected specimens of epiphytes).

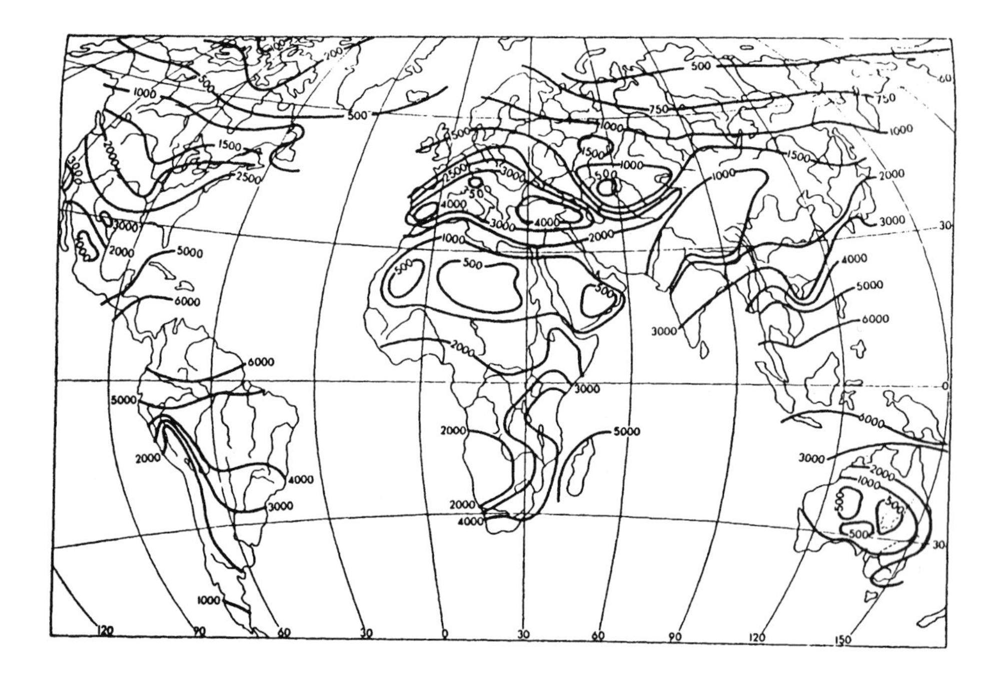

Figure 6. The species diversity of vascular plants on a global scale (from Malyschew 1975).

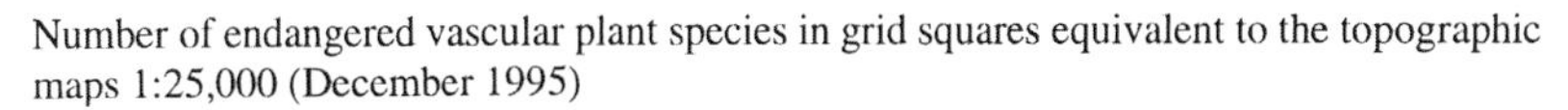
Number of endangered vascular plant species in grid squares equivalent to the topographic maps 1:25,000 (December 1995)

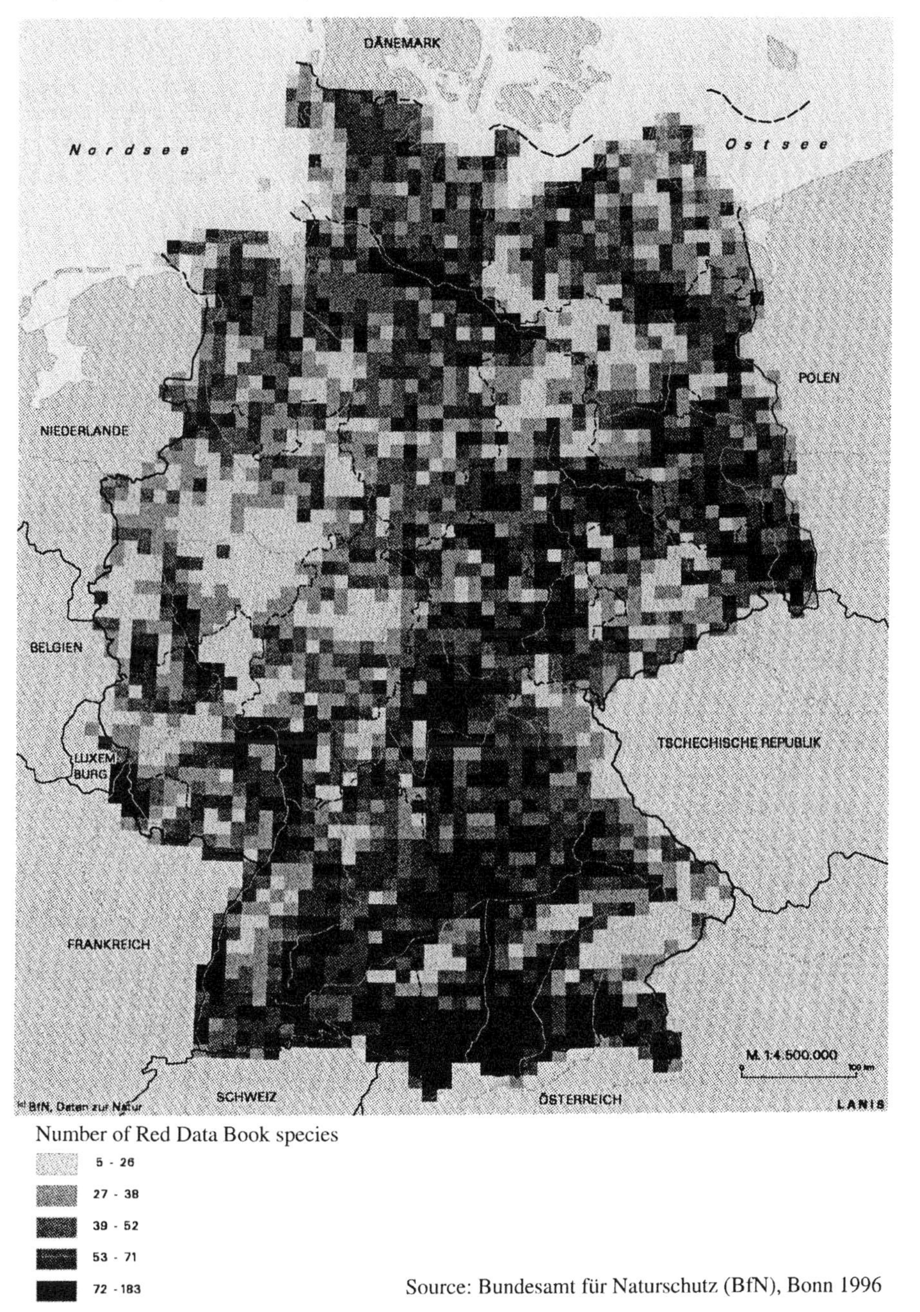

Source: Bundesamt für Naturschutz (BfN), Bonn 1996

Figure 7. Number of endangered and rare vascular plants in Germany in grid squares equivalent to the topographic maps 1:25,000 (that means approximately 11x11 km). Published first time in Haeupler (1997b).

Conclusions

A new thinking on biodiversity can be recognized in some recent literature (see above). However, reliable censuses (like Wisskirchen & Haeupler 1997) and concrete studies in the field (like Seine 1996 and Ibisch 1996) are still largely missing.

In every discussion on biodiversity it should never be forgotten to indicate the level of biodiversity studied (see Fig. 2). Biodiversity per se is a nebulous and ambiguous term. The sample size investigated should be great enough and the scope of the organismic level regarded should not be restricted to a too specific extract. That means, if phytodiversity is studied, all vascular plants or all mosses or all lichens etc. in the investigated ecosystem, area or region should be taken into account.

Neither must the quality of the studied objects ever be neglected. The study of Ibisch (1996) has shown very clearly and proven conclusively that only ecological qualities determine diversity and not qualities like area, number of elements etc.

From this new thinking on biodiversity the conclusion should be drawn that a new funding of biodiversity studies in the field should start. Theoretical thinking about the matter has filled too many pages on glossy paper up to now.

References

Akeroyd, J. 1996. What really is biodiversity? Plant Talk 4: 2.

Barthlott, W., Lauer, W. & Placke, A. in press. Global distribution of species diversity in vascular plants: towards a world map of phytodiversity. Erdkunde 50.

Cailleux, A. 1969. Biogeógraphie mondiale. Presse Universitaire de France, Paris.

Connor, E.F. & McCoy, E.D. 1979. The statistics and biology of the species-area relationship. American Nat. 113: 791-833.

Courrier, K. (ed) 1992. Global Biodiversity Strategy - Guidelines for Action to Save, Study, and Use Earth's Biotic Wealth Sustainably and Equitably. IUCN, Gland.

Duelli, P. 1995. Biodiversität erhalten und fördern: Was sagt die ökologische Forschung dazu? Forum für Wissen, Birmensdorf: 13-21.

Elton, C.S. 1958. The Ecology of Invasions by Animals and Plants. Chapman & Hall, London.

Fisher, R.A., Corbet, A.S. & Williams, C.B. 1943. The relation between the number of species and the number of individuals in a random sample of an animal population. J. Anim. Ecology 12: 42-58.

Frankel, O.H., Brown, A.H.D. & Burdon, J.D. 1995. The Conservation of Plant Biodiversity. Cambridge Univ. Press, Cambridge.

Gaston, K.J. (ed) 1996. Biodiversity. A Biology of Numbers and Difference. Blackwell, Cambridge (USA) etc.

Groombridge, B. (ed) 1992. Global Biodiversity - Status of Earth's Living Resources. A Report Compiled by the World Conservation Monitoring Centre. Chapman & Hall, London etc.

Haeupler, H. 1982. Evenness als Ausdruck der Vielfalt in der Vegetation - Untersuchungen zum Diversitätsbegriff. J. Cramer, Vaduz, Diss. Bot. 65: 1-268.

Haeupler, H. 1995. Diversität. pp. 99-104. In: Kuttler, W. (ed), Handbuch zur Ökologie. Analytica-Verlag, Berlin.

Haeupler, H. 1997a: Islands and Species-Area Curves, a Critical Approach. 36th International Symposium of IAVS, Tenerife 1993. Proceeding Book La Laguna, Serie Informes 40: 141-156.

Haeupler, H. 1997b: Zur Phytodiversitaet Deutschlands: Ein Baustein zu globalen Biodiversitaetsbilanz. Osnabr. Naturwiss. Mitt. 23: 123-133.

Haeupler, H. & Schönfelder, P. 1989. Atlas der Farn- und Blütenpflanzen der Bundesrepublik Deutschland. Ulmer, Stuttgart.

Hawksworth, D.L. (ed) 1995. Biodiversity - Measurement and Estimation. Chapman & Hall, London etc.

Heywood, V.H. & Watson, R.T. (eds) 1995. Global Biodiversity Assessment. Cambridge Univ. Press, Cambridge.

Huston, M.A. 1994. Biological Diversity - The Coexistence of Species on Changing Landscapes. Cambridge Univ. Press, Cambridge.
Ibisch, P.L. 1996. Neotropische Epiphytendiversität - das Beispiel Bolivien. Martina Galunder-Verlag, Wiehl.
ICBP 1992. Putting Biodiversity on the Map: Priority Areas for Global Conservation. Internat. Council for Bird Preservation (ICBP), Cambridge.
Kheong, Y.S. & Win, L.S. 1992. In harmony with nature. Proceedings of the International Conference on Conservation of Tropical Biodiversity. Malayan Nature Society, Kuala Lumpur.
Kim, K.C. & Weaver, R.D. 1995. Biodiversity and Landscapes - A Paradox of Humanity. Cambridge Univ. Press, Cambridge.
Krattinger, A.F., McNeely, J.A., Lesser, W.H., Miller, K.R., St. Hill, Y. & Senanayake, R. (eds) 1994. Widening Perspectives on Biodiversity. IUCN, Gland.
Magurran, A.E. 1988. Ecological Diversity and its Measurement. Princeton Univ. Press, Princeton.
Malyschew, L.J. 1975. The quantitative analysis of flora: Spatial diversity, the level of specific richness and representativity of sampling areas (In Russian). Botan. Zhurnal 60: 1537-1550.
Matthies, D., Schmid, B. & Schmid-Hempel, P. 1995. The importance of population process for the maintenance of biological diversity. GAIA 4: 199-209.
May, R.M. 1992. Wie viele Arten von Lebewesen gibt es? Spektrum der Wissenschaft 12: 72-79.
McArthur, R.H. 1955. Fluctuations of animal populations, and a measure of community stability. Ecology 36: 533-536.
McNeely, J.A., Miller, K.A., Reid, W.V., Mittermeier, R.A. & Werner, T.B. 1990. Conserving the World's Biological Diversity. IUCN, Gland, Washington.
Merxmüller, H. 1966. Systematische Botanik - damals und heute. Ber. Bayer. Bot. Ges. 39: 7-16.
Perring, F.H. & Walters, S.M. 1962. Atlas of the British Flora. Nelson, London.
Ricklefs, R.E. & Schluter, D. 1993. Species Diversity in Ecological Communities - Historical and Geographical Perspectives. Univ. of Chicago Press, Chicago, London.
Schroeder, F.-G. 1969. Zur Klassifizierung der Anthropochoren. Vegetatio 16: 225-238.
Schulze, E.-D. & Mooney, H.A. (eds) 1994. Biodiversity and Ecosystem Function. Springer, Berlin etc.
Seine, R. 1996. Vegetation von Inselbergen in Zimbabwe. Martina Galunder-Verlag, Wiehl.
Senanayake, R. 1994. The need for the measurement and evaluation of biodiversity. pp. 371-375. In: Krattinger *et al.* (eds), Widening Perspectives on Biodiversity. IUCN, Gland.
Signor, P.W. 1990. The geologic history of diversity. Annu. Rev. Ecol. Syst. 21: 509-539.
Simpson, E.H. 1949. Measurement of diversity. Nature 163: 688.
Solbrig, O.T. 1991. Biodiversität - Wissenschaftliche Fragen und Vorschläge für die internationale Forschung. Deutsches Nationalkomitee für das UNESCO-Programm "Man and Biosphere", Bonn.
Sriwatanapongse, S. 1994. Biodiversity conservation and utilisation: Thailand's experience. pp. 171-177. In: Krattinger *et al.* (eds), Widening Perspectives on Biodiversity. IUCN, Gland.
Vitousek, P.M., Loope, L.L. & Adsersen, H. (eds) 1996. Islands. Biological diversity and ecosystem function. Springer, Berlin etc. Ecological Studies 115: 1-238.
Whittaker, R.H. 1972. Evolution and measurement of species diversity. Taxon 21: 213-251.
Williams, C.B. 1964. Patterns in the Balance of Nature. Academic Press, London, New York.
Wilson, E.O. (ed) 1986. Biodiversity. National Academy Press, Washington.
Woodwell, G.M. & Smith, H.H. (eds) 1969. Diversity and Stability in Ecological Systems. Report of Brookhaven Symposium. Biol. Departm. Brookh. Nat. Laboratory. Upton, New York.

BIODIVERSITY - ITS LEVELS AND RELEVANCE FOR NATURE CONSERVATION IN GERMANY

JOSEF BLAB, MANFRED KLEIN & AXEL SSYMANK

Federal Agency for Nature Conservation, Institute for Biotope Protection and Landscape Ecology, Konstantinstr. 110, D-53179 Bonn, Germany

Key words: biodiversity, diversity of biotopes and ecosystems, genetic diversity, nature conservation, nature management, species diversity

Abstract

Since the ratification of the "International Convention of the United Nations on Biodiversity", a lot of activities have been initiated to incorporate biodiversity concepts in nature conservation on national level. While nature conservation in earlier times mainly focused on species conservation, modern strategies are based on the conservation of biodiversity as a whole. This new orientation process is reflected by the "Habitats Directive of the European Union" and the "Biodiversity Convention" itself. Relevance and evaluation of biodiversity, as well as the possibilities for its maintenance are discussed in detail with reference to its main levels of hierarchy: genetic diversity, species diversity, and diversity of habitats or biotopes. The latter includes functional and dynamic aspects of biotopes, biotope complexes and landscape units.

Conservation and management of nature should focus on typical or representative variants of biotopes and ecosystems; this is just as important as protecting rare or threatened species. A Red Data Book of endangered biotopes including a comprehensive list of biotope types occurring in Germany has been compiled. In order to preserve biodiversity, definitions and general guidelines for nature conservation at a regional level must be formulated. Rigid concepts of preserving distinct sites are not sufficient to guarantee the preservation of biodiversity, as dynamic processes and local variability of biotic conditions play an important role. Thus at the same time clear guidelines have to be drawn up for landusers, especially agriculture, forestry and infrastructure.

Introduction

There are many concepts and definitions of diversity in theoretical biosciences (see for example Whittaker 1972). Some of them, like species diversity, are commonly used both in applied vegetation science and in zoology. Biodiversity as a more holistic approach is

A. Kratochwil (ed.), Biodiversity in ecosystems, 199-214

a very recent development in nature conservation, as well as in many adjacent sectors of biological sciences and in ethical, economic and political fields of activity.

Obviously this was strengthened by the UNCED (United Nations Conference for the Environment and Development) in Rio de Janeiro in the year 1992, when the convention on biodiversity was adopted. It was the starting point of numerous activities in the field of conserving and maintaining biodiversity. The convention itself is meanwhile binding for a lot of states, including Germany. With this convention biodiversity has been introduced into public discussions and political hearings all over the world, *e.g.* into the annual conferences of the parties. Of course biodiversity is a very complex subject and there are many definitions and many different understandings of what the biodiversity convention aims at. The common biodiversity view is restricted to pure species diversity, often having in mind mainly tropical rain forests.

Biodiversity is of course the focal point of nature conservation. Nobody will deny the fact that biodiversity as a whole is at risk to an alerting extent. Practical nature conservation cannot be limited to a mere description of the biodiversity of different ecosystems. Based on existing knowledge it must also develop strategies and action plans for its comprehensive maintenance. We want to show the basic significance and possibilities of maintaining biodiversity at our Central European level.

Definition and levels of biodiversity

Concepts of biodiversity in nature conservation

Biodiversity can be defined as the characteristic to be different from one another in a highly specific way which holds true for all living beings (as far as they are not clonal), their habitats, and the system of functional interrelations on all levels of organisation. Biodiversity is represented in two ways (according to Solbrig 1991): A structural and functional aspect in the complexity of morphology and specific functional interrelations on the one hand; second a dynamic aspect within space and time in the specific variability of biological systems under different environmental conditions.

As biological life itself can be regarded as being organised on different hierarchical levels, diversity follows these different levels of organisation (Table 1). But even if biodiversity can be discerned on different hierarchical levels it is quite obvious that these levels do not act independently. A change on one of the levels also implies specific changes on the others. To give one example of the complexity of the system in practical nature conservation: If genetically differentiated populations of a species are preserved, this involves at the same time the conservation of that particular species. But a purely species-orientated preservation strategy, on the other hand, does not necessarily ensure that the genetic diversity of that species can be maintained as well.

Moreover spatial components of biodiversity are of very great importance in conservation evaluations, be it from a global, European, regional, or a very local view. Biodiversity, not being simply a question of diversity of different forms of life (Noah's ark principle), implies also the diversity of functional processes in nature. Thus biodiversity itself cannot be looked upon as being static, but has to be seen as a factor within dynamic systems. Temporary interactions, successional and evolutionary processes play an important role. The conservation of these processes has to be one of the main targets of biodiversity conservation. The generally observed loss of biodiversity occurs on all levels of biodiversity: on the genetic level, on the species level, as well as on the level of ecosystems and biotopes (see UNEP 1995). But the loss of species has attracted most attention. Due to practical and methodical reasons it was easiest to detect, and most of the present literature and discussions still deal with the decline in species numbers. Much less notice has been taken of hidden changes in the genetic diversity as well as of the changes on the more complex levels of biotopes, biotope complexes and ecosystems during the last decades; these should be incorporated to a greater extent in nature conservation policies. We will give a short introduction into the different levels of biodiversity and highlight some of the specific problems that occur on each of the levels. Our principal concern will be ecological, biogeographical and landscape planning aspects of biodiversity in nature conservation.

Table 1. Levels of biological diversity

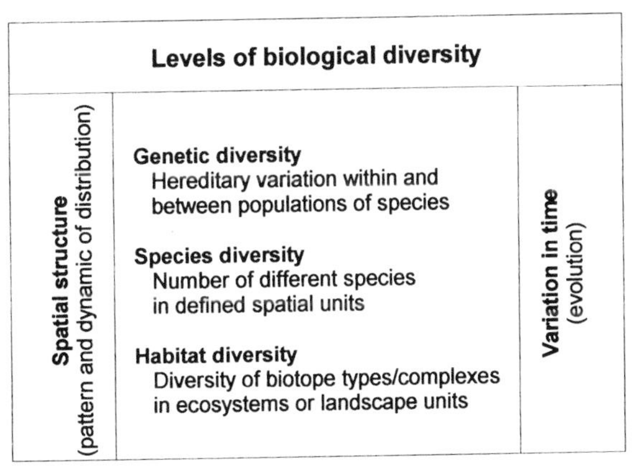

Levels of biological diversity		
Spatial structure (pattern and dynamic of distribution)	**Genetic diversity** Hereditary variation within and between populations of species **Species diversity** Number of different species in defined spatial units **Habitat diversity** Diversity of biotope types/complexes in ecosystems or landscape units	**Variation in time** (evolution)

Genetic diversity

The root of life is mutation-based genetic diversity. Without mutation or genetic diversity, neither life nor variation would exist on earth as all living organisms and biological processes are results of evolution; and evolution is dependent on differences in the genetic information.

According to scientific knowledge the appearance of new mutations is purely a random process. The set of genes of each single organism represents the result of a long and complex process of selection. Thus with the extinction of a particular genetic character also all processes of its creation and selection are lost. Vida (1994) describes evolution as a fascinating process which remembers every success, but forgets all failures.

This is the basic meaning of genetic diversity for biodiversity on higher levels. For practical nature conservation genetic diversity on population level is of special interest. First, evolution acts on population level. Second population is the central unit of consideration for aspects or concepts of

- insulation of landscapes (with effects of isolation/fragmentation)
- Minimum Viable Population (MVP)
- metapopulation dynamics (see Table 2).

Table 2. Concepts in landscape ecology referring to population biology

Concept
Island biogeography theory with species-area-relationship The species number of an island is the result of balanced extinctions and colonisations (turn-over) dependent on island area and distance to the source mainland
Insulation of landscape Increasing isolation of natural, near-natural, or extensively used biotopes due to i.e. traffic arteries, sealing and paving of soil, settlements with related infrastructure, and changes and intensification of agricultural and forestal uses. Its results are fragmentation and increase in ecological resistance for communities and species, with the exception of those adapted to man-made cultural conditions
Minimum viable population Prediction of minimal individual number of an isolated population at a particular site necessary to maintain it with given probability for a given time span
Metapopulation Network of local populations in patchy environments at landscape level interconnected by dispersing propagules (with balance of extinction and recolonisation)

Before presenting some results of our experimental work on genetic diversity in the context of practical biotope conservation, some words about the theoretical frame.

In a very homogeneous and predictable environment the one adaptation to a particular genotype would be optimal that best fits the ruling conditions. In consequence this would mean narrowing genetic diversity within the population. But in a mostly unpredictable and variable environment it is important to be able to react to changes. To keep this ability genetic diversity is an essential raw material.

Like extinct species being irretrievable, lost genetic variants are very unlikely to arise again (even with gene-technological methods). This means that genetic diversity is an unrenewable resource on our planet. In nature a species in general is divided into numerous populations of different sizes, harbouring parts of total genetic diversity. If genetic diversity of species is clustered according to geographic distribution, the units can be named subspecies or races.

During evolutionary times each species (or race) could optimise its strategy to meet environmental demands by selection. But nowadays unprecedented severe human impacts with changes of total landscapes are taking effect on populations and their interactions. Especially habitat reductions or habitat losses as well as isolation and fragmentation should be mentioned when diminishing populations within the remaining natural habitats are studied. But small local populations generally run a higher risk of extinction. Mainly genetic drift and inbreeding entail a reduction of genetic diversity in small local populations.

Modern population-genetic methods have opened possibilities to analyse genetic variation in natural populations. With molecular approaches, for example DNA analysis or the often used isozyme analysis, it is possible to examine hereditary traits acting early in the chain of DNA to phenotypes. Variation can be estimated by the detection of different variants at the same gene locus (polymorphism), the proportion of polymorphic loci, and the numbers of these variants (allele number). Moreover for diploid species the proportion of individuals with different alleles at a polymorphic locus (heterozygosity) can be determined. These parameters are commonly used as measures of genetic variation.

By genetic analyses of three populations of the weevil *Rhinocyllus conicus* (Col.; Curculionidae) the trend of reduced genetic variation in isolated sites can be demonstrated (Table 3).

Table 3. Parameters of genetic diversity of three populations of the weevil *Rhinocyllus conicus* in Rheinland-Pfalz (south-west Germany) (data from Klein 1991)

population	biotope complex	polymorphic loci	allele number	private alleles	hetero-zygosity
Südpfalz	meadow	4	12	1	0.17
Rheinhessen	vineyard fallow	4	10	-	0.09
Soonwald	meadow	5	14	3	0.14

This weevil lives oligophagously in Asteraceae (subtribe Carduineae) and therefore is also used as a biocontrol agent against thistles (Klein 1991). The Rheinhessen population is from a fallow surrounded by intensively managed vineyards dominating the whole landscape; the two other populations have originated from the margins of extensively used meadows. The relatively isolated population from Rheinhessen shows the lowest level of heterozygosity, as well as the lowest number of alleles. The other two populations are genetically more diverse, also containing exclusive own alleles.

However, conservation measures themselves may affect biodiversity on the genetic level, as shown in our analysis of *Isocolus scabiosae* (Cynipidae, Cynipinae, Aulacini). In calcareous grassland biotopes this gall wasp is known as a monophagous gall former of the knapweed *Centaurea scabiosa*. Its galls normally occur at host plant shoots. But in the Eifel region in Germany root galls induced at bottom level were also detected. Emerging individuals of such a root-gall population and two adjacent shoot-gall populations were compared. Since they showed no morphometric differences standard horizontal electrophoresis with cellulose-acetate gels was used. A fixed allele difference was found in the locus GPDH2: all root-gall individuals shared the slow allele whereas all shoot-gall individuals shared the fast one (Table 4). Thus within the species the two genotypes are specialized in quite different plant structures for galling and also are expected to have different phenologies. But both types may be affected differently by mowing or sheep grazing of calcareous grasslands in autumn. (These are the dominant conservation management measures in this region.) As the univoltine gall wasps hibernate as pupae inside the galls, mainly shoot-gall individuals are destroyed, whereas root-gall individuals may survive. This is an example for the thesis that nature conservation measures on a large scale and at the same time may not only reduce the number, but also the genetic diversity of typical biotope-connected populations. Further studies should focus on questions of adaptation and sympatric speciation within this species.

Table 4. GPDH allele frequency of three populations of the gall wasp *Isocolus scabiosae* from the Eifel (western Germany)

Population (n)	root (13)	shoot I (34)	shoot II (24)
Allele A	1.00	0.00	0.00
Allele B	0.00	1.00	1.00

For many species a metapopulation structure is assumed. Adaptations to the special environment within local populations may contribute to a comparably high genetic diversity of total metapopulations and to lower effective population sizes (Gilpin 1991). For example in local populations of the fire salamander (*Salamandra salamandra* L.) allele frequencies of two loci showed marked variations over time. But seen on metapopulation level genetic diversity proved to be more or less constant (Fig. 1).

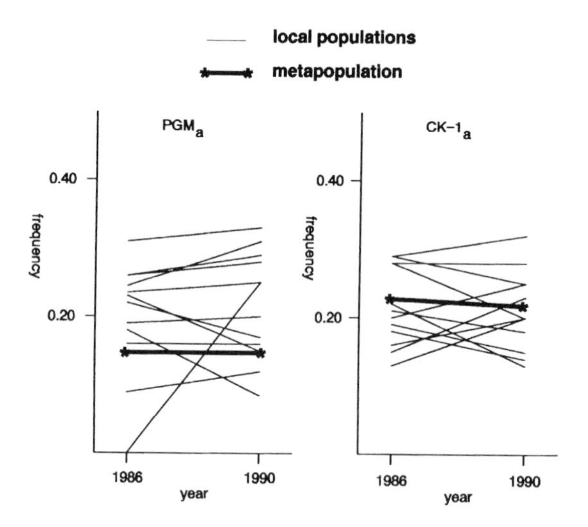

Figure 1. Frequency change of two alleles in populations of *Salamandra salamandra* in the Hunsrück Mountains after four years (modified from Kuhn *et al.* in press).

To meet environmental demands on adaptability, permanently new combinations of the elements of genetic diversity are essential. This means exchange of reproducing individuals (= genes) to keep up connections between local populations. Thus the total area of a metapopulation includes, apart from presently occupied patches, all vacant patches suited for colonisation. In addition the surrounding landscape has to be considered as a matrix for all processes of migration or dispersal. Within a working interacting system of a metapopulation, ecological distance (geographic distance, biotope resistance) between local populations must not prevent gene flow or (re-)colonisation of empty or new patches. Moreover, in most metapopulation concepts central populations are of special importance as they provide surplus individuals for either recolonisation or population support (rescue effect) of adjacent biotopes.

But rules to judge and value changes over time and space, and to differentiate natural versus anthropogenic processes are still lacking. For practical nature conservation, the goal of preservation of genetic diversity has to be defined together with appropriate measures. Does it mean preservation of particular existing geno- or phenotypes (*i.e.* ex situ-conservation, zoos, botanical gardens or gene banks) or does it mean a conservation of the adaptability for survival in natural distribution areas? Ultimately nature conservation aims at keeping natural processes working, like adaptations and evolutionary progress including speciation. These are really long-term processes all based more or less on genetic diversity which in our days may be dramatically reduced in the near future.

Before discussing species diversity two more facets of genetic diversity related to conservation practice should be mentioned (at least briefly). One - exceeding the traditional conservation of natural populations - is the big variety within bred plant and animal species. Here also a threatened reservoir of unique genetic information exists. In former times it was essential to keep robust landraces able to reproduce under prevailing climate, soil, and management methods. Therefore such units are often well-adapted to the very complex conditions of their specific regional environment. However, these aspects are of decreasing importance in modern agriculture and modern breeding. On the other hand the results of modern breeding (high-yield varieties or races) exhibit only na row tolerance thresholds and are dependent on standardisation in cultivation/production methods and - moreover - on high inputs (for example grain: generous application of fertilizers, biocides, and growth retarders). In conservation, especially in traditional cultural landscapes, a lot of questions are linked to the preservation of appropriate regional types of bred animals and plants. Examples are the preservation of traditional orchards or conservation of calcareous grasslands, which both developed in adaptation to the prevailing grazing regime of local stock landraces. Compared to more "modern-bred" stock they are characterized by different grazing selectivity and lower weight.

The second facet concerns isolation of genetic material to gain genes with special properties (*i.e.* genes for resistance or constituents) from locally well-adapted populations. Such isolated genes are to be transferred to breeds of the same species and increasingly also across species borders. However progress in genetic engineering has to keep in mind that genetic diversity of organisms is its most important resource. But the potential risks of genetically modified organisms for the natural ecosystems can only be estimated to a yet very limited extent (Council of Europe 1995). In order to be better able to assess these risks, sound accompanying long-term ecological studies and legally binding regulations on a global scale are required.

Species diversity

Species diversity is mostly used in ecological literature as the number of species present in a particular spatial unit. This may be an administrative unit or simply any other site delimited on a topographic map. Another possibility is to give the number of species of distinct types of habitats or, in phytosociology, of distinct plant associations. This is the easiest mathematical approach to species diversity. A lot of indexes however take relative species abundance into account (see Southwood 1978).

Apart from this, methodical problems of estimating species richness by simply counting species in a set of samples are often enormous, influenced inevitably by the sample number, by the time the samples were taken (during day or during season), as well as by the used methods. Preservation of total species diversity is the declared objective of nature conservation. However, diversity as an abstract term cannot be used in the practical discussion with other competing landusers without valuing ecological functions of certain spatial units. Species diversity alone does not necessarily express stability of ecosystems. To just measure, not to evaluate the latter is an essential tool in the conservation of biotopes.

Thus nature conservation requires both the development of clear methods to measure species diversity and the evaluation of the facts afterwards for nature conservation management. The main goal of nature conservation is to maintain all species and their communities within their natural and semi-natural habitats. This includes spatial diversity due to selection processes as a result of the historical and cultural framework. Maintaining diversity does not mean pure preservationism but a continuation of evolutionary processes which ultimately may lead to speciation.

As long as the habitat or biotope type remain unchanged, a higher species diversity often is positive, indicating better quality (see Fig. 2). On the other hand an increase in species diversity can be the result of negative human influence or pressure on natural systems. An example are new plantations of hedgerow systems, which often include much more plant species than originally present in the particular region (Reif & Aulig 1993). The result may be a change in the species composition of the hedgerow, providing resources and habitats for more specialized phytophagous insects for instance. Succession processes like the ones in abandoned semidry grasslands usually are also accompanied by a clear rise in species diversity in the early succession phases. But this intermediate phase quickly changes to the reverse when shrub species invade the place completely or certain aggressive grass species gain dominance.

Another phenomenon are urban ecosystems often characterized by high diversity indices due to the immigration of many common species or neophytes. These are not at all endangered but present almost everywhere in "unsaturated" ecosystems. These considerations show very clearly that species diversity itself needs a comparative valuation to be useful in nature conservation. Species diversity for its own sake or maximising species numbers of one distinct site or biotope type are not at all sufficient but even may be rather contraproductive. In this context the knowledge of natural regions and characteristic coenoses of a biotope type are very important.

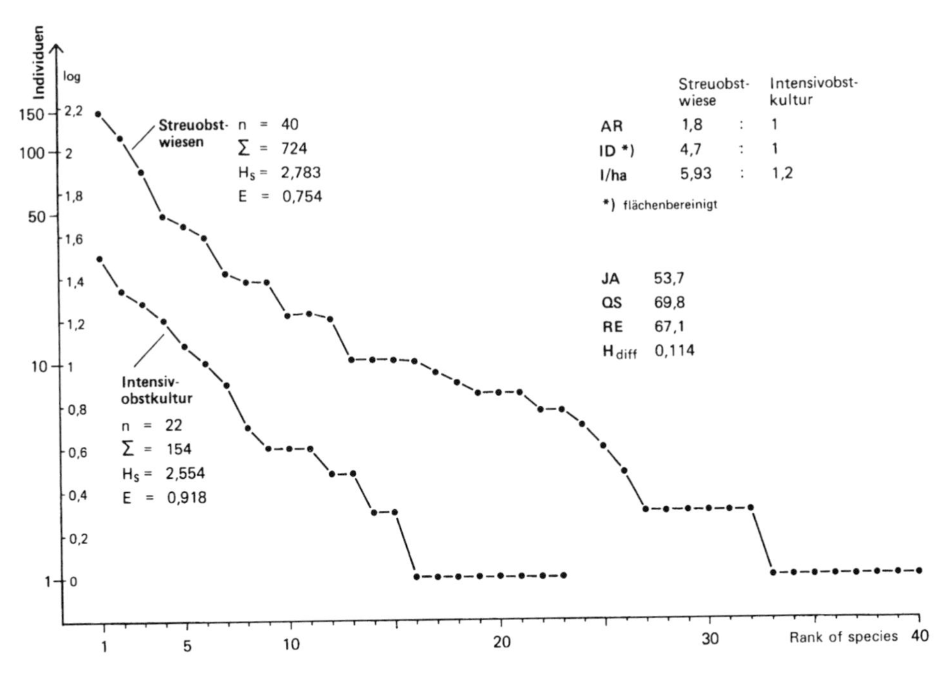

Figure 2. Comparison of avicoenoses of the biotope types traditional extensive orchards ("Streuobstwiesen") and intensive orchards ("Intensivobstkultur") of the Drachenfelser Ländchen near Bonn in summer (according to Blab 1990).

n = number of species
Hs = diversity
AR = relation of species
JA = species diversity after Jaccard 1928
QS = similarity quotient of Sörensen 1948
RE = Renkonen-Index
H_{diff} = diversity difference after MacArthur 1965

S = number of individuals
E = evenness
ID = density of individuals

Criteria for the evaluation or assessment of species diversity which are currently in use are for example:

Degree of threat:

- expressed by the percentage of species listed in Red Data Books, often differentiated according to the degree of threat.

Representativity:

- species typical of a certain natural region, depending on its geographical, pedological and climatic situation.
- species characteristic for the biotope type, considering semiquantitative parameters like frequency or dominance.
- structure and phytosociology, for example number of characteristic species.

Indicator species groups and indicator potential:

- Depending on the specific evaluation necessary for nature management, indicator species or groups of indicator species can be used to evaluate the diversity of ecosystems.

Examples are wetland species assessing the importance of humid pasture systems: whether they occur in the pastures themselves or whether there only is a potential of re-establishing species-rich ecosystems if the plants and animals are still present along ditch systems.

- Percentage of dead-wood indicator species in forest systems using perhaps taxa like macrofungi on wood or certain wood-inhabiting invertebrate groups like Cerambycidae, Coleoptera, different Diptera groups etc. Number of orchid species to assess the diversity of dry and semidry calcareous grasslands. The potential itself can also be evaluated using the species diversity in the seed bank and other indirect techniques.

Another example may illustrate the difficulty of assessing the species diversity of ecosystems. Some ecosystems are very poor in plant species and appear to be very monotonous or even monospecific, like some *Puccinellia* salt meadows or *Phragmites* reed beds, or dune systems dominated by *Ammophila arenaria,* to name but a few. On the other hand these key species often harbour a large series of highly specialized invertebrate groups which may be endophytophagous, epigeic or live in soil, dependent on a certain microstructure only present in this habitat type. Thus an assessment of the species diversity necessitates a very profound analysis of many varying groups of different taxa. At least different functional groups of animals and plant species have to be analysed.

Diversity of biotopes and ecosystems

Biodiversity on the level of coenoses and biotopes requires landscapes with diverse and changing structures, as well as changing ecological conditions. Any modelling of landscapes or landuse practices combined with uniforming or levelling reduce biotope diversity and are in general adverse to nature conservation.

All the scientific knowledge on the level of coenology, especially in zoology, is still at its very beginning. On the other hand there are many different sciences which deal with phytocoenoses: phytosociology, symphenology, syndynamics, synchorology, synevolution etc. Coenoses and their biotopes can be classified into hierarchical systems because they have basic characteristic patterns of organisation and function. Again mathematical approaches, for instance the medium number of species of a plant association, give only a very rough description of the situation. In nature conservation and biotope evaluation characteristic and essential abiotic and biotic conditions and the resulting characteristic combination of species (character species, guilds etc.) are much more important.

Within the framework of a “Red Data Book of Biotopes” for Germany, the “Federal Agency for Nature Conservation” presented a standardised classification of types present in Germany. This list includes about 750 different types classified on different hierarchical levels (Riecken *et al.* 1993). Nevertheless each of those types exhibits specific regional subtypes according to the natural conditions and historical development of the different regions which contribute significantly to the biodiversity in Germany. Other differentiating factors are the presence of

- altitudinal variance, *i.e.* specific subsets of characteristic species dependent on altitude;
- structural variance, *i.e.* different subsets both in vertical and horizontal distribution

patterns. They may be characterized by different dominance structures of guilds;
- micro-variations due to small differences in abiotic conditions or in landuse practice.

Other very important aspects are the cultural and historical situation, and the time available for the development of coenoses and biotope types.

In Central Europe there is literally no square metre which has not, at least for some time, been influenced by man, many areas are actually still under pressure. This is true for most of the forest systems commonly regarded as relatively near-natural habitats.

Depending on the species' ability to migrate and to colonize new habitats, coenoses, especially in new ecosystems, are slowly changing until they reach a quasi-equilibrium. Then the problem arises that some of the species themselves slowly alter their own abiotic conditions and thus induce succession series according to different biotope types.

The highest "level of diversity" and organisation is the diversity of landscapes or of parts of landscapes. Biotopes are organised in distinct patterns, so-called "biotope complexes". They may be visible in zonations along gradients of biotic or abiotic factors or in constant mosaics of spatial distribution. There are multifunctional interdependencies between them, which are especially important for animal species using different habitat types during different phases of their life cycle or for different functions. Juvenile stages for example may live in a completely different habitat type than the adults, or animals may breed in different habitat types, need special overwintering habitats, special habitats for feeding or for raising their youngsters.

According to the species' specific capability of migration and their ability to cross habitat types which normally are not occupied or utilized and may even have adverse effects, completely different spatial distribution patterns may form within a landscape. An example is given in Fig. 3 for flower flies.

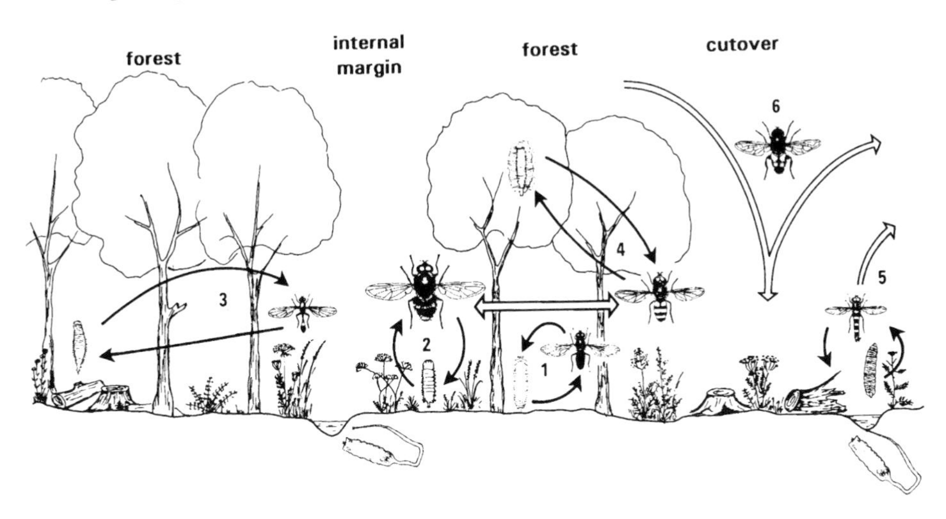

Figure 3. Schematic survey of habitat requirements for flower flies (Syrphidae, Diptera) (modified from Ssymank 1991, according to Blab 1993).

The range covers species which live both as adults and larvae in the same habitat and do not move very far from needed specific structures, up to highly migratory species using flowers as energy source on their migration path.

The diversity of whole landscapes is dependent on many factors, not only on geological/geomorphological and climatic conditions, but, especially in Central Europe, on the history of vegetation development, the history of cultural development, and the actual cultural and socio-economic situation. An example may be the valley of the Oder with its singularities of large alluvial biotope complexes modified severely by anthropogenic use (Fig. 4).

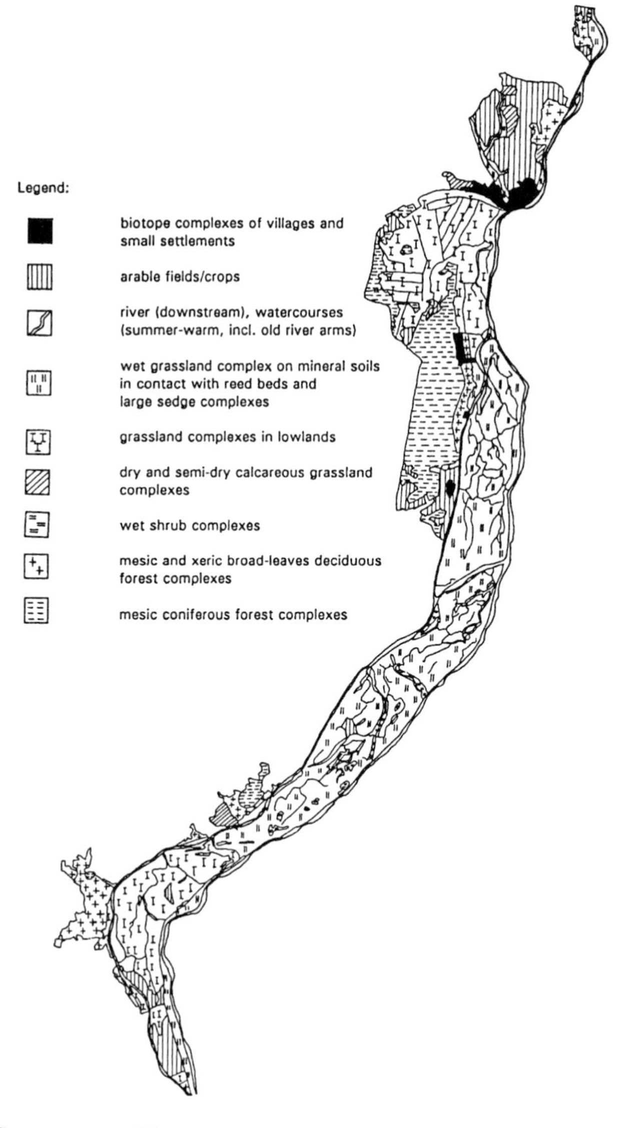

Figure 4. Current pattern of biotope complexes in the Lower Oder valley/north-east Germany (modified from Blab *et al.* 1995b).

The original vegetation of the Oder valley with its broadleaved alluvial forest of *Quercus, Ulmus* and *Alnus* has almost completely disappeared, except for a few remnants near Frankfurt/Oder. But alluvial dynamics and old river arms are still present, furthermore large transition zones between alluvial wetlands and subcontinental semidry sandy meadows or continental steppe meadows of the valley slopes. For these reasons, the Oder valley has to be seen today as a cultural landscape of extraordinary importance for conservation and with a very high diversity of biotopes.

The introduction of biodiversity concepts into practical nature conservation and landuse planning - Provisions and necessities

It is quite obvious that laws and conventions will not be sufficient to maintain biodiversity, as problems arise with efficient practical implementation. Thus we will try to give some basic statements and theses in order to support the development of useful practical concepts for the maintenance and development of biodiversity in nature conservation.

Maintaining biodiversity is only possible with well-defined, uniform and practicable methods for both assessment and evaluation. This has to include for example criteria for setting priorities, minimum requirements, and development or restoration necessities. But also guidelines for any agricultural and forest landuse, respecting environmental demands as well as nature conservation have to be incorporated.

Especially on the level of biotope complexes and landscapes, the scientific basis is rather poor. This is illustrated by the fact that scientific efforts and actual nature conservation policies focus mainly on very rare, endemic or highly threatened species. However, at the same time changes and threats on landscape level are increasing, requiring priorities for sound but quick nature management activities. One important basis for the maintenance of biodiversity is profound knowledge of typical species and species structure of biocoenoses of all existing biotope types and biotope complexes. A prerequisite for assessing the representativity of biotopes and biotope complexes in order to maintain their biodiversity is a detailed knowledge of specific characteristics of biocoenoses in different regional landscape patterns. This is only possible by comparing and analysing different landscapes and establishing data collections for every natural region or subregion.

For Germany we want to maintain our existing habitat diversity over its complete biogeographic range, covering all its variants specific to one or more natural regions. One possibility to achieve this is the establishment of a federal system of priority areas for nature conservation, indicating under scientific aspects the main essential sites for species and biotope conservation. This would open up new ways of landscape planning, aiming at a minimum of conflict situations between nature conservation and other landuse activities.

The “Habitats Directive of the European Union” (Council Directive 92/43/EEC of 21 May 1992 on the conservation of natural habitats and of wild fauna and flora) offers a strong legal framework to maintain biodiversity by establishing a coherent ecological system of protected areas called NATURA 2000. Furthermore it requires the management of specific landscape features which are of major importance for fauna and flora, like stepping stones for migration and dispersal etc. Elements of controlling the efficiency of the measures taken at the particular site have been legally introduced, claiming regular reports on the conservation status every six years. The success of the implementation of this Directive will depend on active measures taken in the member states themselves.

From a pan-European point of view Germany is especially responsible for the conservation and maintenance of some typical and often unique Central European ecosystems: for example the wadden sea, the shallow water zones of the Baltic Sea ("Bodden") with their specific oligohaline conditions, the systems of the northern calcareous Alps and all the large typical beech forest systems (on silicious and calcareous substrates and in a wide altitudinal variation), as well as large Central European river ecosystems (Elbe, Oder) with their remnants of alluvial forests. But also some typical cultural landscapes like semidry pasture systems or hedgerows on stone or earth walls belong to the precious Central European natural heritage.

Biodiversity as a result of historical and evolutionary processes cannot be generated at will. For the maintenance of biodiversity a much higher "value" has to be conceded to natural and semi-natural features and ecosystems. The possibility of a restoration of habitats and ecosystems has to be taken into account much more in landuse planning. Biotope types and complexes where even a long-term restoration is impossible or will occur only in minor parts, cannot constantly be overridden by so-called public or economic interests. Ecosystems are no technical buildings or constructions that can be transplanted or rebuilt after destruction, even if necessary technical and financial resources were available. The claim of land, for whatever purpose, has to be restricted by an impact assessment firmly respecting the needs for conservation and maintenance of biodiversity; temporarily strong economic or public interests must not be the decisive factor when decisions about land claims are made.

The "Federal Agency for Nature Conservation" thus established a "Red Data Book for Biotope Types and Biotope Complexes" (Riecken *et al.* 1993). It was the result of long and detailed analyses and experts judgements, taking into account both quantitative loss in area as well as qualitative changes in two parallel categories of threat (Blab *et al.* 1995b). At the same time the regeneration ability of biotopes, being partly destroyed or locally extinct, was assessed for each of the biotope types present in Germany. This will be a valuable tool for a better objective impact assessment, also in the framework of the necessary compensation, according to the Federal law for nature conservation. Generally environmental impact studies will be, due to their accuracy and safety of judgement, of special benefit, as most of the actual studies still consider only certain species.

Nature conservation strategies in general should be consistently adapted to regional and local situations, and consequently a federal conservation approach to landscape development has to work out regionalized models, which define principal aims for maintaining biodiversity (Finck *et al.* 1993). Such concepts should be based on a thorough investigation and assessment of the biotic and abiotic status quo, the natural potential of an area, as well as on man's use of the land in its historic development up to the present day. A development model is ultimately the result of balancing the ideal conditions from a federal conservation view against the demands of the respective socio-economic situation.

Landuse policies have to respect nature conservation in order to guarantee a long-term wise use essential for maintenance of biodiversity. A respective nature-friendly adapted intensity of agriculture and forestry will not only safeguard the biological diversity of cultivated species or crops, but also help to manage and maintain intermingled or adjacent remnants of more natural and valuable biotopes and biotope complexes. Taking into account the need to preserve and develop the variety of our cultural landscapes, crops, species and coenosis mode and intensity of use have to be attuned much more to the ecological demands of the particular site. This may in some cases oblige to maintain non-profitable landuse forms on a limited scale, including traditional races of stock and crops.

Biotope protection has to be regarded as much more important, especially in the regional context of typical landscape features. Of course protection in this sense does not at all mean pure preservation, but to respect more and more also natural processes of dynamics and variability. The aim of nature conservation must be to protect species, populations and biotopes not only in static systems of clearly delimited sites, but within a functional system of the landscape they live in. This includes possibilities for the development of biotope types or populations in new localities in a dynamic equilibrium, while otherwise such sites may vanish. But of course nowadays the limits of such a system are always getting narrower with increasing pressure on landscape by competing landusers.

References

Blab, J. 1990. Zum Indikationspotential von Roten Listen und zur Frage der Ermittlung "Regionaler Leitartengruppen" mit landschaftsökologischer Zeigerfunktion. Schriftenreihe für Landschaftspflege und Naturschutz 32: 121-134.

Blab, J. 1993. Grundlagen des Biotopschutzes für Tiere. 4th ed. Kilda-Verlag, Greven.

Blab, J., Klein, M. & Ssymank, A. 1995a. Biodiversität und ihre Bedeutung in der Naturschutzarbeit. Natur und Landschaft 70: 11-18.

Blab, J., Riecken, U. & Ssymank, A. 1995b. Proposal on a criteria system of a National Red Data Book of Biotopes. Landscape Ecology 10: 41-50.

Council of Europe (ed) 1995. Pan-European conference on the potential long-term ecological impact of genetically modified organisms, proceedings. Environmental Encounters, No. 20, Council of Europe Press, Strasbourg.

Finck, P., Hauke, U. & Schröder, E. 1993. Zur Problematik der Formulierung regionaler Landschafts-Leitbilder aus naturschutzfachlicher Sicht. Natur und Landschaft 68: 603-607.

Gilpin, M.E. 1991. The genetic effective size of a metapopulation. pp. 165-175. In: Gilpin, M.E. & Hanski, I. (eds), Metapopulation Dynamics: Empirical and Theoretical Investigations. Academic Press, London.

Klein, M. 1991. Populationsbiologische Untersuchungen an *Rhinocyllus conicus* FRÖLICH (Col.; Curculionidae). Wiss.-Verlag Maraun, Frankfurt.

Kuhn, J., Veith, M. & Klein M. in press. Das Metapopulationskonzept im Amphibienschutz. Zeitschrift für Ökologie und Naturschutz.

Riecken, U., Ries, U. & Ssymank, A. 1993. Biotoptypenverzeichnis für die Bundesrepublik Deutschland. Schriftenreihe für Landschaftspflege und Naturschutz 38: 301-339.

Riecken U., Ries, U. & Ssymank, A. 1994. Rote Liste der gefährdeten Biotoptypen der Bundesrepublik Deutschland. Schriftenreihe für Landschaftspflege und Naturschutz 41.

Reif, A. & Aulig, G. 1993. Künstliche Neupflanzung naturnaher Hecken - Sinnvolle Naturschutztechnologie oder unlösbarer Widerspruch? Naturschutz und Landschaftsplanung 25: 85-93.

Southwood, T.R.E. 1978. Ecological Methods. 2nd ed. Chapman & Hall, London.

Tasks for vegetation science

1. E.O. Box: *Macroclimate and plant forms.* An introduction to predictive modelling in phytogeography. 1981 ISBN 90-6193-941-0
2. D.N. Sen and K.S. Rajpurohit (eds.): *Contributions to the ecology of halophytes.* 1982 ISBN 90-6193-942-9
3. J. Ross: *The radiation regime and architecture of plant stands.* 1981 ISBN 90-6193-607-1
4. N.S. Margaris and H.A. Mooney (eds.): *Components of productivity of Mediterranean-climate regions.* Basic and applied aspects. 1981 ISBN 90-6193-944-5
5. M.J. Müller: *Selected climatic data for a global set of standard stations for vegetation science.* 1982 ISBN 90-6193-945-3
6. I. Roth: *Stratification of tropical forests as seen in leaf structure* [Part 1]. 1984 ISBN 90-6193-946-1
   For Part 2, see Volume 21
7. L. Steubing and H.-J. Jäger (eds.): *Monitoring of air pollutants by plants.* Methods and problems. 1982 ISBN 90-6193-947-X
8. H.J. Teas (ed.): *Biology and ecology of mangroves.* 1983 ISBN 90-6193-948-8
9. H.J. Teas (ed.): *Physiology and management of mangroves.* 1984 ISBN 90-6193-949-6
10. E. Feoli, M. Lagonegro and L. Orláci: *Information analysis of vegetation data.* 1984 ISBN 90-6193-950-X
11. Z. Šesták (ed.): *Photosynthesis during leaf development.* 1985 ISBN 90-6193-951-8
12. E. Medina, H.A. Mooney and C. Vázquez-Yánes (eds.): *Physiological ecology of plants of the wet tropics.* 1984 ISBN 90-6193-952-6
13. N.S. Margaris, M. Arianoustou-Faraggitaki and W.C. Oechel (eds.): *Being alive on land.* 1984 ISBN 90-6193-953-4
14. D.O. Hall, N. Myers and N.S. Margaris (eds.): *Economics of ecosystem management.* 1985 ISBN 90-6193-505-9
15. A. Estrada and Th.H. Fleming (eds.): *Frugivores and seed dispersal.* 1986 ISBN 90-6193-543-1
16. B. Dell, A.J.M. Hopkins and B.B. Lamont (eds.): *Resilience in Mediterranean-type ecosystems.* 1986 ISBN 90-6193-579-2
17. I. Roth: *Stratification of a tropical forest as seen in dispersal types.* 1987 ISBN 90-6193-613-6
18. H.-G. Dässler and S. Börtitz (eds.): *Air pollution and its influence on vegetation.* Causes, Effects, Prophylaxis and Therapy. 1988 ISBN 90-6193-619-5
19. R.L. Specht (ed.): *Mediterranean-type ecosystems.* A data source book. 1988 ISBN 90-6193-652-7
20. L.F. Huenneke and H.A. Mooney (eds.): *Grassland structure and function.* California annual grassland. 1989 ISBN 90-6193-659-4
21. B. Rollet, Ch. Högermann and I. Roth: *Stratification of tropical forests as seen in leaf structure,* Part 2. 1990 ISBN 0-7923-0397-0
22. J. Rozema and J.A.C. Verkleij (eds.): *Ecological responses to environmental stresses.* 1991 ISBN 0-7923-0762-3
23. S.C. Pandeya and H. Lieth: *Ecology of Cenchrus grass complex.* Environmental conditions and population differences in Western India. 1993 ISBN 0-7923-0768-2
24. P.L. Nimis and T.J. Crovello (eds.): *Quantitative approaches to phytogeography.* 1991 ISBN 0-7923-0795-X
25. D.F. Whigham, R.E. Good and K. Kvet (eds.): *Wetland ecology and management.* Case studies. 1990 ISBN 0-7923-0893-X

Tasks for vegetation science

26. K. Falinska: *Plant demography in vegetation succession.* 1991 ISBN 0-7923-1060-8
27. H. Lieth and A.A. Al Masoom (eds.): *Towards the rational use of high salinity tolerant plants,* Vol. 1: Deliberations about high salinity tolerant plants and ecosystems. 1993 ISBN 0-7923-1865-X
28. H. Lieth and A.A. Al Masoom (eds.): *Towards the rational use of high salinity tolerant plants,* Vol. 2: Agriculture and forestry under marginal soil water conditions. 1993 ISBN 0-7923-1866-8
29. J.G. Boonman: *East Africa's grasses and fodders.* Their ecology and husbandry. 1993 ISBN 0-7923-1867-6
30. H. Lieth and M. Lohmann (eds.): *Restoration of tropical forest ecosystems.* 1993 ISBN 0-7923-1945-1
31. M. Arianoutsou and R.H. Groves (eds.): *Plant-animal interactions in Mediterranean-type ecosystems.* 1994 ISBN 0-7923-2470-6
32. V.R. Squires and A.T. Ayoub (eds.): *Halophytes as a resource for livestock and for rehabilitation of degraded lands.* 1994 ISBN 0-7923-2664-4

KLUWER ACADEMIC PUBLISHERS – DORDRECHT / BOSTON / LONDON